T0310823

A First Course in Probability for Computer and Data Science

A First Course in Probability for Computer and Data Science

Henk Tijms

Vrije University, The Netherlands

World Scientific

NEW JERSEY · LONDON · SINGAPORE · BEIJING · SHANGHAI · HONG KONG · TAIPEI · CHENNAI · TOKYO

Published by

World Scientific Publishing Co. Pte. Ltd.

5 Toh Tuck Link, Singapore 596224

USA office: 27 Warren Street, Suite 401-402, Hackensack, NJ 07601

UK office: 57 Shelton Street, Covent Garden, London WC2H 9HE

Library of Congress Cataloging-in-Publication Data

Names: Tijms, H. C., author.

Title: A first course in probability for computer and data science / Henk Tijms.

Description: New Jersey : World Scientific, [2023] | Includes index.

Identifiers: LCCN 2023013602 | ISBN 9789811271748 (hardcover) |
    ISBN 9789811272042 (paperback) | ISBN 9789811271755 (ebook)

Subjects: LCSH: Probabilities. | Computer science--Mathematics. | Data mining--Statistical methods.

Classification: LCC QA273 .T474 2023 | DDC 004.01/51--dc23/eng/20230605

LC record available at https://lccn.loc.gov/2023013602

British Library Cataloguing-in-Publication Data

A catalogue record for this book is available from the British Library.

For any available supplementary material, please visit
https://www.worldscientific.com/worldscibooks/10.1142/13291#t=suppl

Printed in Singapore

Preface

Probability is the bedrock for data analysis and statistics. Students in computer and data science need a solid background in probability, especially for areas such as machine learning and artificial intelligence. This undergraduate text is a truly accessible introduction to the fundamental principles of probability. The book emphasizes probabilistic and computational thinking rather than theorems and proofs. It provides insights and motivates the students by showing them why probability works and how to apply it.

The book starts at the beginning; no specific knowledge of probability is required. It introduces probabilistic ideas and probability models with their solution methods that are most useful in computer and data science. Unique features of this undergraduate textbook are the Bayesian approach to inference, the interaction between probability and Monte Carlo simulation, real-world applications of probability, Poisson heuristic for weakly dependent trials, and a gentle introduction to Markov chains.

In Chapter 1 the basics of combinatorial analysis are discussed along with the important role of the exponential function in probability. Chapter 2 covers not only standard material such as sample space, conditional probability, discrete random variables, expected value, standard deviation, the square-root law, the law of large numbers, and generating functions, but it also covers several topics not found in most introductory texts, such as Bayesian probability with real-life cases in law and medicine, naive Bayes and logistic regression, and the Kelly strategy for gambling and investing. Chapter 3 deals with specific probability distributions that are useful in applications.

First, the binomial, hypergeometric, and Poisson distributions are discussed and the connection between them. Attention is also given to the important Poissonization method for the multinomial probability experiment. Next, the most important continuous distributions are introduced. The normal probability distribution and the central limit theorem are discussed in detail. Much attention is also given to the exponential distribution and the Poisson process. The Q-Q plot and the chi-square test are also covered. The chapter is concluded with a brief discussion of the bivariate normal distribution. Chapter 4 covers ten real-world applications of probability. Insight into the occurrence of coincidences in lotteries and birthday-type of problems are given by using the Poisson distribution. Benford's law and coupon collecting are also discussed. Chapter 5 highlights the role of Monte Carlo simulation in probability. Teaching probability and statistics well is not easy. Monte Carlo methods may be very helpful. The basic tools of Monte Carlo simulation are treated and illustrated with many examples. Attention is also given to the statistical analysis of simulation output, including confidence intervals for the simulated estimates. Chapter 6 gives a primer on Markov chains. The core idea of Markov chains is state and state transition, which is extremely useful for both modeling and computational purposes. Many probability problems can be solved by an appropriately chosen Markov chain. Chapter 6 also presents a lucid introduction to Markov chain Monte Carlo simulation and discusses both the Metropolis-Hastings algorithm and the Gibbs sampler. All chapters are interspersed with historical facts about probability.

The text contains many worked examples. Problems are an integral part of the text. Many instructive problems scattered throughout the text are given along with problem-solving strategies. Several of the problems extend previously covered material. Answers to all problems and worked-out solutions to selected problems are provided.

Contents

Chapter 1

Combinatorics and a Few Calculus Facts

This chapter presents a number of results from combinatorics and calculus, in preparation for the subsequent chapters. Section 1.1 introduces you to the concepts of factorials and binomial coefficients. In Section 1.2 the exponential function and the natural logarithm will be discussed.

1.1 Combinatorial analysis

Many probability problems require counting techniques. In particular, these techniques are extremely useful for computing probabilities in a chance experiment in which all possible outcomes are equally likely. In such experiments, one needs effective methods to count the number of outcomes in any specific event. In counting problems, it is important to know whether the order in which the elements are counted is relevant or not. Factorials and binomial coefficients will be discussed and illustrated.

In the discussion below, the *fundamental principle of counting* is frequently used: if there are a ways to do one activity and b ways to do another activity, then there are $a \times b$ ways of doing both. As an example, suppose that you go to a restaurant to get some breakfast. The menu says pancakes, waffles, or fried eggs, while for a drink you can choose between juice, coffee, tea, and hot chocolate. Then the total number of different choices of food and drink is $3 \times 4 = 12$. As another example, how many different license plates are possible

1

when the license plate displays a nonzero digit, followed by three letters, followed by three digits? The answer is that the total number of possible license plates is

$$9 \times 26 \times 26 \times 26 \times 10 \times 10 \times 10 = 158\,184\,000.$$

Example 1.1. How many ordinary five-card poker hands containing four of a kind are possible?

Solution. You can choose the four cards of a kind $(A, K,\ldots, 2)$ in 13 ways. The fifth card can be chosen in 48 ways. Thus, $13 \times 48 = 624$ ordinary five-card poker hands containing four of a kind are possible.

Factorials and permutations

How many different ways can you order a number of different objects such as letters or numbers? For example, what is the number of different ways that the three letters A, B, and C can be ordered? By writing out all the possibilities ABC, ACB, BAC, BCA, CAB, and CBA, you can see that the total number is 6. This brute-force method of writing down all the possibilities and counting them is naturally not practical when the number of possibilities gets large, as is the case for the number of possible orderings of the 26 letters of the alphabet. You can also determine that the three letters A, B, and C can be ordered in 6 different ways by reasoning as follows. For the first position, there are 3 available letters to choose from, for the second position, there are 2 letters left over to choose from, and only one letter for the third position. Therefore, the total number of possibilities is $3 \times 2 \times 1 = 6$. The general rule should now be evident. Suppose that you have n distinguishable objects. How many ordered arrangements of these objects are possible? Any ordered sequence of the objects is called a *permutation*. Reasoning in the same way as above gives that there are n ways for choosing the first object, leaving $n - 1$ choices for the second object, etc. Therefore, the total number of ways to order n *distinguishable* objects is equal to the product $n \times (n - 1) \times \cdots \times 2 \times 1$. This product is denoted by $n!$ and is called 'n factorial'. Thus, for any positive integer n,

$$\boxed{n! = 1 \times 2 \times \cdots \times (n - 1) \times n.}$$

A useful convention is

$$0! = 1,$$

which simplifies the presentation of several formulas to be given below. Note that $n! = n \times (n-1)!$ and so $n!$ grows very quickly as n gets larger. For example, $5! = 120$, $10! = 3\,628\,800$ and $15! = 1\,307\,674\,368\,000$. Summarizing, for any positive integer n,

the total number of ordered sequences (permutations) of n distinguishable objects is $n!$.

Example 1.2. Eight important heads of state, including the U.S. President and the British Premier, are present at a summit conference. For the perfunctory group photo, the eight dignitaries are lined up next to one other. What is the number of possible arrangements in which the U.S. President and the British Premier stand next to each other?

Solution. If the positions of the U.S. President and the British Premier are fixed, there remain $6!$ possible arrangements for the other six statesmen. The U.S. President and the British Premier stand next to each other if they take up the positions i and $i+1$ for some i with $1 \leq i \leq 7$. If these two statesmen take up the positions i and $i+1$, there are $2!$ possibilities for the order among them. Thus, the total number of possible arrangements in which the U.S. President and the British Premier stand next to each other is $6! \times 7 \times 2! = 10\,080$.

Example 1.3. How many different words can be composed from 11 letters consisting of five letters A, two letters B, two letters R, one letter C, and one letter D?

Solution. Imagine that the five letters A are numbered as A_1 to A_5, the two letters B as B_1 and B_2, and the two letters R as R_1 and R_2. Then you have 11 different letters, and the number of ways to order those letters is $11!$. The five letters A_1 to A_5, the two letters B_1 and B_2, and the two letters R_1 and R_2 can among themselves be ordered in $5! \times 2! \times 2!$ ways. Each of these orderings gives the same word. Thus, the total number of different words that can be formed from the original 11 letters is

$$\frac{11!}{5! \times 2! \times 2!} = 83\,160.$$

Thus the word ABRACADABRA will appear with probability $\frac{1}{83\,160}$ when the 11 letters are put in random order.

Binomial coefficients and combinations

How many different juries of three persons can be formed from five persons A, B, C, D, and E? By direct enumeration, you see that the answer is 10: $\{A, B, C\}$, $\{A, B, D\}$, $\{A, B, E\}$, $\{A, C, D\}$, $\{A, C, E\}$, $\{A, D, E\}$, $\{B, C, D\}$, $\{B, C, E\}$, $\{B, D, E\}$, and $\{C, D, E\}$. In this problem, the order in which the jury members are chosen is not relevant. The answer 10 juries could also have been obtained by a basic principle of counting. First, count how many juries of three persons are possible when attention is paid to the order. Then determine how often each group of three persons has been counted. Thus, the reasoning is as follows. There are 5 ways to select the first jury member, 4 ways to then select the next member, and 3 ways to select the final member. This would give $5 \times 4 \times 3$ ways of forming the jury when the order in which the members are chosen would be relevant. However, this order makes no difference. For example, for the jury consisting of the persons A, B, and C, it is not relevant which of the 3! ordered sequences ABC, ACB, BAC, BCA, CAB, and CBA has led to the jury. Hence, the total number of ways a jury of 3 persons can be formed from a group of 5 persons is equal to $\frac{5 \times 4 \times 3}{3!}$. This expression can be rewritten as

$$\frac{5 \times 4 \times 3 \times 2 \times 1}{3! \times 2!} = \frac{5!}{3! \times 2!}.$$

In general, you can calculate that the total number of possible ways to choose a jury of k persons out of a group of n persons is equal to

$$\frac{n \times (n-1) \times \cdots \times (n-k+1)}{k!}$$
$$= \frac{n \times (n-1) \times \cdots \times (n-k+1) \times (n-k) \times \cdots \times 1}{k! \times (n-k)!} = \frac{n!}{k! \times (n-k)!}.$$

This leads to the definition

$$\boxed{\binom{n}{k} = \frac{n!}{k! \times (n-k)!}}$$

for non-negative integers n and k with $k \leq n$. The quantity $\binom{n}{k}$ (pronounce: n choose k) has the interpretation:

> $\binom{n}{k}$ **is the total number of ways to choose k different objects out of n distinguishable objects, paying no attention to their order.**

In other words, $\binom{n}{k}$ is the total number of combinations of k different objects out of n and is referred to as the *binomial coefficient*. The key difference between permutations and combinations is *order*. Combinations are *unordered* selections, permutations are *ordered* arrangements.

The binomial coefficients play a key role in the so-called *urn model*. This model has many applications in probability. Suppose that an urn contains R red and W white balls. What is the probability of getting exactly r red balls when blindly grasping n balls from the urn? To answer this question, it is helpful to imagine that the balls are made distinguishable by giving each of them a different label.[1] The total number of possible combinations of n different balls is $\binom{R+W}{n}$. Under these combinations there are $\binom{R}{r} \times \binom{W}{n-r}$ combinations with exactly r red balls (and thus $n-r$ white balls). Thus, if you blindly grasp n balls from the urn, then

$$\boxed{\text{the probability of getting exactly } r \text{ red balls} = \frac{\binom{R}{r} \times \binom{W}{n-r}}{\binom{R+W}{n}}}$$

with the convention that $\binom{a}{b} = 0$ for $b > a$. These probabilities represent the so-called *hypergeometric distribution*. Probability problems that can be translated into the urn model appear in many disguises. A nice illustration is the lottery 6/45. In each drawing of the lottery,

---

[1] Labeling objects to distinguish them from each other can be very helpful when solving a combinatorial probability problem.

six different numbers are chosen from the numbers $1, 2, \ldots, 45$. Suppose you have filled in one ticket with six distinct numbers. Then the probability of matching exactly r of the drawn six numbers is

$$\frac{\binom{6}{r} \times \binom{39}{6-r}}{\binom{45}{6}} \quad \text{for } r = 0, 1, \ldots, 6,$$

as you can see by identifying the six drawn numbers with 6 red balls and the other 39 numbers with 39 white balls. In particular, the probability of matching all six drawn numbers (the jackpot) equals 1 to 8 145 060.

Example 1.4. How many ordinary five-card poker hands containing one pair are possible? One pair means two cards of the same face value plus three cards with different face values.

Solution. The answer to the question requires careful counting to avoid double counting. To count the number of hands with one pair, you proceed as follows. Pick the face value for the pair: $\binom{13}{1}$ choices. Pick two cards from the face value: $\binom{4}{2}$ choices. Pick three other face values: $\binom{12}{3}$ choices. Pick one card from each of the other three face values: $4 \times 4 \times 4$ choices. This gives that the total number of possible hands with one pair is

$$\binom{13}{1} \times \binom{4}{2} \times \binom{12}{3} \times 4^3 = 1\,098\,240.$$

You can choose five cards out of the 52 playing cards in $\binom{52}{5}$ ways, and so the probability of getting a five-card poker hand with one pair is

$$\frac{1\,098\,240}{\binom{52}{5}} = 0.4226.$$

Example 1.5. Six socks are lost when washing ten different pairs of socks. How many combinations of seven matching pairs are possible for the remaining socks? How many combinations are possible for four matching pairs?

Solution. You are left with seven complete pairs of socks only if both socks of three pairs are missing. You can choose three pairs in

$\binom{10}{3} = 120$ ways. Thus, 120 combinations of seven complete pairs are possible for the remaining socks. You are left with four matching pairs of socks only if exactly one sock of each of six pairs is missing. These six pairs can be chosen in $\binom{10}{6}$ ways. There are two possibilities for how to choose one sock from a given pair. This means that $\binom{10}{6} \times 2^6 = 13\,440$ combinations of four matching pairs are possible for the remaining socks. It is much more likely that four matching pairs of socks remain than seven matching pairs of socks. The probability of seven complete pairs and the probability of four complete pairs have the values

$$\frac{120}{\binom{20}{6}} = \frac{1}{323} \quad \text{and} \quad \frac{13\,440}{\binom{20}{6}} = \frac{112}{323}.$$

When things go wrong, they really go wrong!

Combinatorial identities

In mathematics there are many identities in which binomial coefficients appear. The following recursive relation is known as Pascal's triangle[2]:

$$\binom{n}{k} = \binom{n-1}{k-1} + \binom{n-1}{k} \quad \text{for } 1 \leq k \leq n.$$

You can algebraically prove this. A more elegant proof is by interpreting the same 'thing' in two different ways. This is called a word-proof. Think of a group of n persons from which a committee of k persons must be chosen. The k persons can be chosen in $\binom{n}{k}$ ways. However, you can also count as follows. Take a particular person, say John. The number of possible committees containing John is given by $\binom{n-1}{k-1}$, and the number of possible committees not containing John is given by $\binom{n-1}{k}$, which verifies the identity.

---

[2] Pascal was far from the first to study this triangle. The Persian mathematician Al-Karaji had produced something very similar as early as the 10th century, and the triangle is called Yang Hui's triangle in China after the 13th century Chinese mathematician Yang Hui, and Tartaglia's triangle in Italy after the 16th century Italian mathematician Niccolò Tartaglia.

Test questions

• How many distinct license plates with three letters followed by three digits are possible? How many if the letters and numbers must be different? (answer: $17\,576\,000$ and $11\,232\,000$)

• What is the total number of ways to arrange 5 letters A and 3 letters B in a row? (answer: 56)

• Five football players A, B, C, D, and E are designated to take a penalty kick after the end of a football match. In how many orders can they shoot if A must shoot immediately after C? How many if A must shoot after C? (answer: 24 and 60)

• What is the total number of distinguishable permutations of the eleven letters in the word Mississippi? (answer: $34\,650$)

• John and Pete are among 10 players who are to be divided into two teams A and B, each consisting of five players. How many formations of the two teams are possible so that John and Pete belong to a same team? (answer: 112)

• Suppose that from 10 children, five are to be chosen and lined up. How many different lines are possible? (answer: $30\,240$)

• How many ordinary five-card poker hands containing two pairs plus one card with a different face value are possible? (answer: $123\,552$)

• Five dots are placed on a 7×7 grid so that no cell contains more than one dot. How many configurations are possible so that no row or column contains more than one dot? (answer: $52\,920$)

• Give word proofs of $\binom{n}{n-k} = \binom{n}{k}$ and $\sum_{k=0}^{n} \binom{n}{k}\binom{n}{n-k} = \binom{2n}{n}$.

• How many ways are there to distribute eight identical chocolate bars between five children so that each child gets at least one chocolate bar? (answer: 35)[3]

---

[3]The number of combinations of *non-negative* integers $x_1, \ldots, x_n$ satisfying $x_1 + \cdots + x_n = r$ is equal to $\binom{n+r-1}{r}$. This result is stated without proof.

1.2 The exponential and logarithmic functions

The history of the number e begins with the discovery of logarithms by John Napier in 1614. At this time in history, international trade was experiencing a period of strong growth, and as a result, there was much attention given to the concept of compound interest. At that time, it was already noticed that $(1 + \frac{1}{n})^n$ tends to a certain limit if n is allowed to increase without bound:

$$\lim_{n \to \infty} \left(1 + \frac{1}{n}\right)^n = e \text{ with } e = 2.7182818\ldots.$$

The famous mathematical constant e is called the Euler number, and it crops up everywhere in the field of probability. This constant is named after Leonhard Euler (1707–1783) who is considered as the most productive mathematician in history.

The *exponential function* is defined by e^x, where the variable x runs through the real numbers. This is one of the most important functions in mathematics. A fundamental property of e^x is that this function has itself as derivative:

$$\frac{de^x}{dx} = e^x \quad \text{for all } x.$$

How to calculate the function e^x? The generally valid relation

$$\lim_{n \to \infty} \left(1 + \frac{x}{n}\right)^n = e^x \quad \text{for all } x$$

is not useful for that purpose. The calculation of e^x is based on

$$e^x = 1 + x + \frac{x^2}{2!} + \frac{x^3}{3!} + \cdots \quad \text{for all } x.$$

The proof of this power series expansion requires Taylor's theorem from calculus. The fact that e^x has itself as derivative is crucial in the proof. Note that term by term differentiation of the series $1 + x + \frac{x^2}{2!} + \cdots$ leads to the same series, in agreement with the fact that e^x has itself as derivative.

The series expansion of e^x leads to $e^x \approx 1+x$ for x close to 0. This is one of the most useful approximation formulas in mathematics! In probability theory the formula is often used as

$$\boxed{e^{-x} \approx 1 - x \quad \text{for } x \text{ close to } 0.}$$

A nice illustration of the usefulness of this formula is provided by the birthday problem. What is the probability that two or more people share a birthday in a randomly formed group of m people (no twins)? To simplify the analysis, it is assumed that the year has 365 days (February 29 is excluded) and that each of these days is equally likely as birthday. Number the people as 1 to m and let the sequence $(v_1, v_2, \ldots, v_m)$ denote their birthdays. The total number of possible sequences is $365 \times 365 \times \cdots \times 365 = 365^m$, while the number of sequences in which each person has a different birthday is $365 \times 364 \times \cdots \times (365 - m + 1)$. Denoting by P_m the probability that each person has a different birthday, you have

$$P_m = \frac{365 \times 364 \times \cdots \times (365 - m + 1)}{365^m}.$$

If m is much smaller than 365, the insightful approximation

$$P_m \approx e^{-\frac{1}{2}m(m-1)/365}$$

applies. To see this, write P_m as

$$P_m = 1 \times \left(1 - \frac{1}{365}\right) \times \left(1 - \frac{2}{365}\right) \times \cdots \times \left(1 - \frac{m-1}{365}\right).$$

Next, by $e^{-x} \approx 1 - x$ for x close to zero and the algebraic formula $1 + 2 + \cdots + n = \frac{1}{2}n(n+1)$ for $n \geq 1$, you get

$$P_m \approx e^{-1/365} \times e^{-2/365} \times \cdots \times e^{-(m-1)/365} = e^{-(1+2+\ldots+m-1)/365}$$
$$= e^{-\frac{1}{2}m(m-1)/365}.$$

The sought probability that two or more people share a same birthday is one minus the probability that each person has a different birthday. Thus,

probability of two or more people sharing a birthday
$$\approx 1 - e^{-\frac{1}{2}m(m-1)/365}.$$

This probability is already more than 50% for $m = 23$ people (the exact value is 0.5073, and the approximate value is 0.5000). The intuitive explanation that the probability of a match is already more than 50% for such a small value as $m = 23$ is that there are $\binom{23}{2} = 253$ combinations of two persons, each combination having a matching probability of $\frac{1}{365}$.

Natural logarithm

The function e^x is strictly increasing on $(-\infty, \infty)$ with $\lim_{x \to -\infty} e^x = 0$ and $\lim_{x \to \infty} e^x = \infty$. Therefore, for each fixed $c > 0$, the equation $e^y = c$ on $(-\infty, \infty)$ has a unique solution y. This solution as function of c is called the *natural logarithm*. It is denoted by $\ln(c)$ for $c > 0$. Thus, the natural logarithm is the inverse function of the exponential function. The function $\ln(x)$ is the logarithmic function with base e. In statistical computing, the relationship

$$e^{\ln(a)} = a \quad \text{for any } a > 0$$

can be very helpful. Logarithms have the property that they enable you to reduce the manipulation with extremely large or extremely small numbers to the manipulation with moderately sized numbers. Doing calculations on a log scale and then exponentiating them usually resolves numerical problems of overflow or underflow.

The natural logarithm can also be defined by the integral

$$\ln(y) = \int_1^y \frac{1}{v}\, dv \quad \text{for } y > 0.$$

This integral representation of $\ln(y)$ shows that

$$\frac{d\ln(y)}{dy} = \frac{1}{y} \quad \text{for } y > 0.$$

The integral formula for $\ln(x)$ implies $\ln(n + 1) \leq H_n \leq 1 + \ln(n)$, where $H_n = 1 + \frac{1}{2} + \cdots + \frac{1}{n}$ is the partial sum of the harmonic series. A very accurate approximation for H_n is

$$1 + \frac{1}{2} + \cdots + \frac{1}{n} \approx \ln(n) + \gamma + \frac{1}{2n},$$

where $\gamma = 0.57721\ldots$ is the Euler-Mascheroni constant. The absolute error of the approximation is bounded by $\frac{1}{8n^2}$.[4]

Geometric series

In probability analysis you will often encounter the geometric series. The basic formula for the geometric series is

$$1 + x + x^2 + \cdots = \frac{1}{1-x} \quad \text{for } |x| < 1,$$

or, shortly, $\sum_{k=0}^{\infty} x^k = \frac{1}{1-x}$ for $|x| < 1$. You can easily verify this result by working out $(1-x)(1 + x + x^2 + \cdots + x^m)$ as $1 - x^{m+1}$. If you take $|x| < 1$ and let m tend to infinity, then x^{m+1} tends to 0. This gives $(1-x)(1 + x + x^2 + \cdots) = 1$ for $|x| < 1$, which verifies the desired result. Differentiating the geometric series term-by-term and noting that $\frac{1}{1-x}$ has $\frac{1}{(1-x)^2}$ as derivative, you get

$$\sum_{k=1}^{\infty} kx^{k-1} = \frac{1}{(1-x)^2} \quad \text{for } |x| < 1.$$

By differentiating both sides of this equation, you get

$$\sum_{k=2}^{\infty} k(k-1)x^{k-2} = \frac{2}{(1-x)^3} \quad \text{for } |x| < 1.$$

Integrating both sides of $\frac{1}{1+x} = 1 - x + x^2 - x^3 + \cdots$ for $|x| < 1$, you get the series expansion

$$\ln(1+y) = y - \frac{y^2}{2} + \frac{y^3}{3} - \frac{y^4}{4} + \cdots \quad \text{for } |y| < 1.$$

---

[4]The harmonic series $\sum_{k=1}^{\infty} \frac{1}{k}$ has the value ∞. There are many proofs for this celebrated result. The first proof dates back to about 1350 and was given by the philosopher Nicolas Oresme. His argument is ingenious. Oresme simply observed that $\frac{1}{3} + \frac{1}{4} > \frac{2}{4} = \frac{1}{2}$, $\frac{1}{5} + \frac{1}{6} + \frac{1}{7} + \frac{1}{8} > \frac{4}{8} = \frac{1}{2}$, $\frac{1}{9} + \frac{1}{10} + \cdots + \frac{1}{16} > \frac{8}{16} = \frac{1}{2}$, etc. In general, $\frac{1}{r+1} + \frac{1}{r+2} + \cdots + \frac{1}{2r} > \frac{1}{2}$ for any r, showing that $\sum_{k=1}^{n} \frac{1}{k}$ eventually grows beyond any bound as n gets larger. Isn't it a beautiful argument?

Chapter 2

Fundamentals of Probability

Probability is the science of uncertainty and it is everywhere:

• What is the chance of winning the jackpot in the national lottery?

• What is the chance of having some rare disease if tested positive?

• What is the chance that the last person to draw a ticket will be the winner if one prize is raffled among 10 people?

• How many cards would you expect to draw from a standard deck before seeing the first ace?

• What is the expected value of your loss when you are going to bet 50 times on red in roulette?

• What is the expected number of different values that come up when six fair dice are rolled? What is the expected number of rolls of a fair die it takes to see all six sides of the die?

The tools to answer these kinds of questions will be given in this chapter, which aims to familiarize yourself with the most important basic concepts in elementary probability. The standard axioms of probability are introduced, the important properties of probability are derived, the key ideas of conditional probability and Bayesian thinking are covered, and the concepts of random variable, expected value, and standard deviation are explained. All this is illustrated with insightful examples and instructive problems.

2.1 Foundation of probability

Approximately four hundred years after the colorful Italian mathematician and physician Gerolamo Cardano (1501–1576) wrote his book *Liber de Ludo Aleae* (*Book on Games of Chance*) and laid a cornerstone for the foundation of the field of probability by introducing the concept of *sample space*, celebrated Russian mathematician Andrey Kolmogorov (1903–1987) cemented that foundation with axioms on which a solid theory can be built.

The sample space of a chance experiment is a set of elements that one-to-one correspond to all of the possible outcomes of the experiment. Here are some examples:

• The experiment is to roll a die once. The sample space can be taken as the set $\{1, 2, \ldots, 6\}$, where the outcome i means that i dots appear on the up face.

• The experiment is to repeatedly roll a die until the first six shows up. The sample space can be taken as the set $\{1, 2, \ldots\}$ of the positive integers. Outcome k indicates that a six appears for the first time on the kth roll.

• The experiment is to measure the time until the first emission of a particle from a radioactive source. The sample space can be taken as the set $(0, \infty)$ of the positive real numbers, where the outcome t means that it takes a time t until the emission of a particle.

In the first example, the sample space is a finite set. In the second example, the sample space is a so-called countably infinite set, while in the third example, the sample space is a so-called uncountable set.

In set theory, a non-finite set is called *countably infinite* if the elements of the set one-to-one correspond to the natural numbers. Not all sets with a non-finite number of elements are countably infinite. The set of all points on a line and the set of all real numbers between 0 and 1 are examples of infinite sets that are not countable. Sets that are neither finite nor countably infinite are called *uncountable*, whereas sets that are either finite or countably infinite are called *countable*.

The idea of Kolmogorov was to consider a sufficiently rich class of subsets of the sample space and to assign a number, $P(A)$, between 0 and 1, to each subset A belonging to this class of subsets. The class of subsets consists of all possible subsets if the sample space is finite or countably infinite, but certain 'weird' subsets must be excluded if the sample space is uncountable. For the probability measure P, three natural postulates are assumed. Denoting by $P(A \text{ or } B)$ the number assigned to the set of all outcomes belonging to either subset A or subset B or to both, the axioms for a finite sample space are:

Axiom 1. $P(\Omega) = 1$ *for the sample space* Ω.

Axiom 2. $0 \le P(A) \le 1$ *for each subset A of* Ω.

Axiom 3. $P(A \text{ or } B) = P(A) + P(B)$ *if the subsets A and B have no element in common (so-called disjoint subsets).*

Axiom 3 needs a modification if the sample space is non-finite. Then the axiom is $P(A_1 \text{ or } A_2 \text{ or } \ldots) = \sum_{k=1}^{\infty} P(A_k)$ for pairwise disjoint subsets $A_1, A_2, \ldots$. The sample space endowed with a probability measure P on the class of subsets is called a *probability space*.

If the sample space contains a countable number of elements, it is sufficient to assign a probability $p(\omega)$ to each element ω of the sample space. The probability $P(A)$ that is then assigned to a subset A of the sample space is defined by the sum of the probabilities of the individual elements of set A. That is, in mathematical notation,

$$P(A) = \sum_{\omega \in A} p(\omega).$$

A special case is the case of a finite sample space in which each outcome is equally likely. Then $P(A)$ can be calculated as

$$P(A) = \frac{\text{the number of outcomes belonging to } A}{\text{the total number of outcomes of the sample space}}.$$

This probability model is known as the *Laplace model*, named after the famous French scientist Pierre Simon Laplace (1749–1827), who is sometimes called the 'French Newton'. What is called the Laplace model was first introduced by Gerolamo Cardano in his 16th century book. This probability model was used to solve a main problem in early probability: the probability of not getting a 1 in two rolls of a fair die is $\frac{25}{36}$. Galileo Galilei (1564–1642), one of the greatest scientists of the Renaissance, used the model to explain to the Grand Duke of Tuscany, his benefactor, that it is more likely to get a sum of 10 than a sum of 9 in a single roll of three fair dice (the probabilities are $\frac{27}{216}$ and $\frac{25}{216}$).

In probability language, any subset A of the sample space is called an *event*. It is said that event A occurs if the outcome of the experiment belongs to the set A. The number $P(A)$ is the probability that event A will occur. Any individual outcome is also an event, but events correspond typically to more than one outcome. For example, the sample space of the experiment of a single roll of a die is the set $\{1, 2, 3, 4, 5, 6\}$, where outcome i means that i dots appear on the up face of the die. Then, the subset $A = \{1, 3, 5\}$ represents the event that an odd number shows up. Events A and B are called *mutually exclusive* (or *disjoint*) if they cannot both occur at the same time.

The probability measure P does not appear out of thin air, rather you must consciously choose it. Naturally, this must be done in such a way that the axioms are satisfied and the model reflects the reality of the problem at hand in the best possible way. The axioms must hold true not only for the interpretation of probabilities in terms of relative frequencies for a repeatable experiment such as the rolling of a die. They must also remain valid for the Bayesian interpretation of probability as a measure of personal belief in the outcome of a non-repeatable experiment, for example, a horse race. A subjective probability depends on one's knowledge or information about the event in question.

Example 2.1. You are randomly dealt four cards from an ordinary deck of 52 playing cards. What is the probability of getting no ace?

Solution. Two methods will be presented to solve this problem. It is always helpful if you can check the solution using alternative solution methods.

Solution method 1: This method uses an unordered sample space. The sample space consists of all possible combinations of four different cards. This sample space has $\binom{52}{4}$ equally likely elements. The number of elements for which there is no ace among the four cards is $\binom{48}{4}$. Thus, the probability of getting no ace is

$$\frac{\binom{48}{4}}{\binom{52}{4}} = 0.7187.$$

Solution method 2: This method uses an ordered sample space. The sample space consists of all possible orderings of the 52 cards. The sample space has 52! equally likely elements. The number of possible orderings for which there is no ace among the first four cards in the ordering is $48 \times 47 \times 46 \times 45 \times 48!$. Thus, the probability of getting no ace can also be found as

$$\frac{48 \times 47 \times 46 \times 45 \times 48!}{52!} = 0.7187.$$

Example 2.2. Three players enter a room and are given a red or a blue hat to wear. The color of each hat is determined by a fair coin toss. Players cannot see the color of their own hats, but do see the color of the other two players' hats. The game is won when at least one of the players correctly guesses the color of his own hat, and no player gives an incorrect answer. In addition to having the opportunity to guess a color, players may also pass. No communication of any kind between players is allowed after they have been given hats; however, they may agree on a group strategy beforehand. The players decided upon the following strategy. A player who sees that the other two players wear a hat with the same color guesses the opposite color for his/her own hat; otherwise, the player says nothing. What is the probability of winning the game under this strategy?

Solution. This problem boils down to the chance experiment of tossing a fair coin three times. As sample space, take the set consisting of the eight elements RRR, RRB, RBR, BRR, BBB, BBR, BRB, and RBB, where R stands for a red hat and B for a blue hat. Each element of the sample space is equally probable and gets assigned a probability of $\frac{1}{8}$. The strategy is winning if one the six outcomes RRB, RBR, BRR, BBR, BRB, or RBB occurs (verify!). Thus, the probability of winning the game under the chosen strategy is $\frac{3}{4}$.

Sample points may be easily incorrectly counted. In his book *Opera Omnia*, the German mathematician Gottfried Wilhelm Leibniz (1646–1716) — inventor of differential and integral calculus along with Isaac Newton — made a famous mistake by stating: "with two dice, it is equally likely to roll twelve points than to roll eleven points because one or the other can be done in only one manner". He argued: two sixes for a sum 12, and a five and a six for a sum 11. However, there are two ways to get a sum 11; that is obvious by imagining that one die is blue and the other is red. Alternatively, you can think of two rolls of a single die instead of a single roll of two dice.

Another psychologically tempting mistake that is sometimes made is to treat sample points as equally likely when this is actually not the case. This mistake can be illustrated with a famous misstep of Jean le Rond d'Alembert (1717–1783) who was one of the foremost intellectuals of his time. D'Alembert made the error to state that the probability of getting heads in no more than two coin tosses is $\frac{2}{3}$ rather than $\frac{3}{4}$. He reasoned as follows: "once heads appears upon the first toss, there is no need for a second toss. The possible outcomes of the game are thus H, TH, and TT, and so the required probability is $\frac{2}{3}$". However, these three outcomes are not equally likely, but should be assigned the probabilities $\frac{1}{2}$, $\frac{1}{4}$, and $\frac{1}{4}$, respectively. The correct answer is $\frac{3}{4}$, as would be immediately clear from the sample space $\{HH, HT, TH, TT\}$ for the experiment of two coin tosses.

Example 2.3. Two desperados, A and B, are playing a game of Russian roulette using a gun. One of the gun's six cylinders contains a bullet. The desperados take turns pointing the gun at their own heads and pulling the trigger. Desperado A begins. If no fatal shot is fired, they give the cylinder a spin such that it stops at a random chamber, and the game continues with desperado B, and so on. What is the probability that desperado A will fire the fatal shot?

Solution. The set $\{1, 2, \ldots\}$ of the positive integers is taken as sample space for the experiment. Outcome i means that the fatal shot occurs at the ith trial. In view of the independence of the trials, an appropriate probability model is constructed by assigning the probability $\frac{1}{6}$ to outcome 1, the probability $\frac{5}{6} \times \frac{1}{6}$ to outcome 2, and so on. In general, the probability assigned to outcome i is

$$p_i = \frac{5}{6} \times \cdots \times \frac{5}{6} \times \frac{1}{6} = \left(\frac{5}{6}\right)^{i-1} \times \frac{1}{6} \quad \text{for } i = 1, 2, \ldots.$$

This is an example of the so-called geometric distribution.[5] Let A be the event that desperado A fires the fatal shot. This event occurs for the outcomes $1, 3, 5, \ldots$. Thus, using the geometric series $\sum_{k=0}^{\infty} x^k = \frac{1}{1-x}$ for $|x| < 1$, you get

$$P(A) = \sum_{k=0}^{\infty} p_{2k+1} = \frac{1}{6} \sum_{k=0}^{\infty} \left(\frac{25}{36}\right)^k = \frac{1}{6} \times \frac{1}{1 - 25/36},$$

and so the probability that desperado A will fire the fatal shot is $\frac{6}{11}$.

Product rule for a compound chance experiment

The chance experiment from Example 2.3 is a so-called *compound chance experiment*. Such an experiment consists of a sequence of elementary sub-experiments. In the compound experiment from Example 2.3, the sub-experiments are physically independent of each other, that is, the outcome of one sub-experiment does not affect the outcome of any other sub-experiment. The probabilities to the outcomes of the compound experiment were assigned by taking the product of probabilities of individual outcomes of the sub-experiments. This is the only assignment that reflects the physical independence of the sub-experiments.

How to assign probabilities to the outcomes of a compound chance experiment when the sub-experiments are not physically independent? This is also done by a *product rule*. To explain this rule, consider the experiment of sequentially picking two balls at random from a box containing four red and two blue balls, where the first picked ball is not put back in the box when drawing the second ball. What is the probability of picking at least one blue ball? This experiment is a compound experiment with two physically dependent sub-experiments. The sample space of the compound experiment consists

---

[5]In general, this is the probability distribution of the number of trials until the first success occurs in a sequence of independent trials each having the same success probability p. A very useful probability model!

of the four ordered pairs (r, r), (r, b), (b, r), and (b, b), where the first component of each pair indicates the color of the first picked ball and the second component indicates the color of the second picked ball. Outcome (r, r) gets assigned the probability $p(r, r) = \frac{4}{6} \times \frac{3}{5} = \frac{2}{5}$. The rationale behind this assignment is that the first ball you pick will be red with probability $\frac{4}{6}$. If the first ball you pick is red, three red and two blue balls remain in the box, in which case the second ball you pick will be red with probability $\frac{3}{5}$. By the same argument, outcome (r, b) gets assigned the probability $p(r, b) = \frac{4}{6} \times \frac{2}{5} = \frac{4}{15}$, $p(b, r) = \frac{2}{6} \times \frac{4}{5} = \frac{4}{15}$, and $p(b, b) = \frac{2}{6} \times \frac{1}{5} = \frac{1}{15}$. This probability model is an adequate representation of the experiment and enables us to answer the question of what the probability of picking at least one blue ball is. This probability is

$$p(r, b) + p(b, r) + p(b, b) = \frac{4}{15} + \frac{4}{15} + \frac{1}{15} = \frac{3}{5}.$$

The basis of the probability model used to obtain this answer was a product rule. This rule will be encountered again in Section 2.3 when discussing conditional probabilities.

It is fun to also give a probability problem with an uncountable sample space.

Example 2.4. The game of franc-carreau was a popular game in eighteenth-century France. In this game, a coin is tossed on a chessboard. The player wins if the coin does not fall on one of the lines of the board. Suppose a coin with a diameter of d is blindly tossed on a large table. The surface of the table is divided into squares whose sides measure a in length such that $a > d$. What is the probability of the coin falling entirely within the confines of a square?

Solution. The trick is to concentrate on the center point of the coin. Take as sample space the square in which this point falls. The meaning of a point in the sample space is that the center of the coin lands on that point. Since the coin lands randomly on the table, the probability that is assigned to each measurable subset A of the sample space is the area of the region A divided by the area of the square. The coin falls entirely within a square if and only if the center

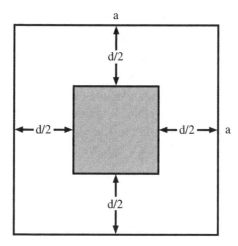

Figure 1: Franc-carreau game.

point of the coin lands on a point in the shaded square in Figure 1. The area of the shaded square is $(a - d)^2$. Therefore,

$$P(\text{the coin will fall entirely within a square}) = \frac{(a - d)^2}{a^2}.$$

Complement rule

One of the most useful calculation rules in the field of probability is the complement rule, which states that the probability of a given event occurring can be found by calculating the probability that the event will not occur. These two probabilities sum to 1. The complement rule is often used to find the probability of 'something' occurring at least once. For example, the rule is very helpful to find the probability of at least one six occurring in four rolls of a die, and the probability of at least one double six in 24 rolls of two dice, see Problem 2.2 below. This probability problem has an interesting history. The French nobleman Chevalier de Méré was a famous gambler of the 17th century. He frequently offered the bet that he could obtain a six in four rolls or less of a single die, and the bet that he could obtain a double six with two dice in 24 rolls or less. The Chevalier believed that the chance of winning the bet was the same in both games (can you explain why the respective chances are not $\frac{4}{6}$ and $\frac{24}{36}$?). In reality, however, he won the first game more often than not. The Chevalier approached the mathematician Blaise Pascal for clarification. This inquiry led to a correspondence between the two famous French mathematicians Blaise Pascal (1623–1662) and Pierre de Fermat (1601–1665).[6] They mathematically clarified the dice problem by simply calculating the chances of *not* rolling a six or double six, see Problem 2.2 below.

The *complement rule* says that, for any event A,

$$\boxed{P(A) = 1 - P(\overline{A}),}$$

where the *complementary event* $\overline{A}$ is defined as the event that A does not occur. A formal proof of this obvious result goes as follows. The events A and $\overline{A}$ are mutually exclusive, and together they form the whole sample space. Then, by the Axioms 1 and 3, you have $P(A \text{ or } \overline{A}) = 1$ and $P(A \text{ or } \overline{A}) = P(A) + P(\overline{A})$.

Example 2.5. You hear that a draw in the 6/45 lotto has produced six consecutive numbers. How unlikely do you think this is?

Solution. The probability that the numbers 1 to 6 will appear in the upcoming draw of the 6/45 lotto is $\binom{6}{6}/\binom{45}{6}$, see Section 1.1. So the probability of six consecutive numbers is $p = 40 \times \binom{6}{6}/\binom{45}{6} = 4.91 \times 10^{-6}$. This is a very small probability, but the picture changes if the question is put into context. There are many lottos in the world. Imagine one hundred 6/45 lottos each with two draws per week. The probability that in the next, say, 5 years, no six consecutive numbers will appear in the $100 \times 5 \times 52 \times 2 = 52\,000$ draws is $(1 - p)^{52\,000} = 0.7746$, by the product rule. The complement rule then gives that the probability of at least one draw occurring with six

---

[6]The 1654 Pascal–Fermat correspondence marks the beginning of modern probability theory. In this correspondence, another famous probability problem was solved. Chevalier de Méré had also brought to Pascal's attention the problem of points, in which the question is how the winnings of a game of chance should be divided between two players if the game was ended prematurely. This problem will be discussed in Section 3.1.

consecutive numbers is $1 - 0.7746 = 0.2254$. This is by no means a small probability. The lesson is that, however improbable an event may be, it will almost certainly occur at some point if the event is given a sufficiently large number of opportunities to manifest itself.

Problem 2.1. A dog has a litter of four puppies. Use an appropriate sample space to verify that it is more likely that the litter consists of three puppies of the same gender and one of the other than two puppies of each gender. (answer: $\frac{8}{16}$ versus $\frac{6}{16}$)

Problem 2.2. Use an appropriate sample space to argue that the probability of getting at least one six in r rolls of a single die is $1 - \frac{5^r}{6^r}$, and the probability of getting at least one double six in r rolls of two dice is $1 - \frac{35^r}{36^r}$. What are the smallest values of r for which the probabilities are more than 0.5? (answer: $r = 4$ and $r = 25$)

Problem 2.3. (a) What is the probability of getting two or more times a same number in one roll of three dice? (answer: $\frac{4}{9}$)
(b) Two dice are rolled. If the biggest number is 1, 2, 3, or 4, player 1 wins; otherwise, player 2. Who has an edge? (answer: player 2)

Problem 2.4. Three balls are randomly placed one by one in three boxes. What is the probability that exactly one box remains empty? (answer: $\frac{2}{3}$)

Problem 2.5. You have two gift cards, each loaded with 10 free drinks from your favorite coffee shop. Each time you get a drink, you randomly pick one of the cards to pay with. One day, it happens for the first time that the waiter can't accept the card because it does not have any drink credits left on it. What is the probability that the other card has also no free drinks on it? (answer: 0.1762)

Problem 2.6. In a game show, a father and his daughter are standing in front of three closed doors, behind which a car, the key to the car, and a goat are hidden in random order. Each of them can open up to two doors, one at a time, and this must be done out of sight of the other. The daughter is given the task of finding the car, and the father must find the key. Only if both are successful, they get to keep the car. Father and daughter are allowed to discuss a strategy before the game starts. What is an optimal strategy? What is the maximum probability of winning the car? (answer: $\frac{4}{6}$)

Problem 2.7. For non-disjoint sets A and B, the *sum rule* is $P(A \text{ or } B) = P(A) + P(B) - P(A \text{ and } B)$, where '$A$ and B' is the set of outcomes belonging to both A and B.[7] Can you explain this rule? What is the probability of getting an ace or a heart when picking randomly one card from a deck of 52 cards? (answer: $\frac{16}{52}$)

Problem 2.8. An experiment has three possible outcomes O_1, O_2, and O_3 with probabilities $p_1 = 0.10$, $p_2 = 0.15$, and $p_3 = 0.75$, respectively. What is the probability that outcome O_1 will appear before O_2 if the experiment is done repeatedly? (answer: 0.40)

Problem 2.9. Two people have agreed to meet at the train station. Independently of one other, each person is to appear at a random moment between 12 p.m. and 1 p.m. What is the probability that they will meet within 10 minutes of each other? (answer: $\frac{11}{36}$)

2.2 The concept of conditional probability

The concept of conditional probability lies at the heart of probability theory. It is an intuitive concept. To illustrate this, most people reason as follows to find the probability of getting two aces when two cards are selected at random in succession from an ordinary deck of 52 cards. The probability of getting an ace on the first card is $\frac{4}{52}$. Given that one ace is gone from the deck, the probability of getting an ace on the second card is $\frac{3}{51}$. Therefore,

$$P(\text{the first two cards are aces}) = \frac{4}{52} \times \frac{3}{51}.$$

What is applied here is the *product rule* for probabilities:

$$\boxed{P(A \text{ and } B) = P(A)P(B \mid A),}$$

---

[7]In mathematical notation, $P(A \text{ or } B)$ is written as $P(A \cup B)$, and $P(A \text{ and } B)$ as $P(A \cap B)$. The sum rule reads as $P(A \cup B) = P(A) + P(B) - P(A \cap B)$ in the set-theoretic notation.

where $P(A$ and $B)$ stands for the probability that both event A ('the first card is an ace') and event B ('the second card is an ace') will occur, $P(B \mid A)$ is the notation for the conditional probability that event B will occur given that event A has occurred.[8] In words, the unconditional probability that both event A and event B will occur is equal to the unconditional probability that event A will occur times the conditional probability that event B will occur given that event A has occurred. This is one of the most useful rules in probability.

Example 2.6. Someone is looking to rent an apartment on the top floor of a certain building. The person gets wind of the fact that two apartments in the building are empty and are up for rent. The building has seven floors with eight apartments per floor. What is the probability of having a vacant apartment on the top floor?

Solution. There are two possible approaches to solving this problem. In both, the complement rule is applied. This means that, instead of calculating the probability in question, you calculate the complementary probability of no top floor apartment being available. Subtracting this probability from 1 gives the probability of having a vacant apartment on the top floor.

Approach 1: This approach is based on counting and requires the specification of a sample space. The elements of the sample space are all possible combinations of two of the 56 apartments. The total

---

[8]In fact, the other way around, $P(B \mid A)$ is defined as the ratio of $P(A$ and $B)$ and $P(A)$ if $P(A) > 0$. This definition can be motivated as follows. Suppose that n physically independent repetitions of a chance experiment are done under the same conditions. Let r be the number of times that event A occurs simultaneously with event B, and s be the number of times that event A occurs but not event B. The frequency at which event B occurs in the cases that event A has occurred is equal to $\frac{r}{r+s}$. The frequency at which both event A and event B occur is $\frac{r}{n}$, and the frequency at which event A occurs is $\frac{r+s}{n}$. The ratio of these frequencies is $\frac{r}{r+s}$. This ratio is exactly the frequency at which event B occurs in the cases that event A has occurred. This explains the definition of $P(B \mid A)$

The conditional probability $P(B \mid A)$ is in fact a probability measure on a reduced sample space. For example, suppose a blue and red die are rolled and you get the information that there is a six among the two outcomes. Then the reduced sample space of the experiment is $\{(1,6), \ldots, (5,6), (6,1), \ldots (6,6)\}$, where outcome (i, j) means that the blue and red die show i and j points.

number of possible combinations is $\binom{56}{2} = 1\,540$, whereas the number of possible combinations without a vacant apartment on the top floor is $\binom{48}{2} = 1\,128$. Then, taking the ratio of all favorable combinations and the total number of combinations,

$$P(\text{no apartment is vacant on the top floor}) = \frac{1\,128}{1\,540} = 0.7325.$$

Approach 2: The second approach is based on conditional probabilities. Imagine that the two available apartments were vacated one after the other. Then, let A be the event that the first vacant apartment is not located on the top floor and B be the event that the second vacant apartment is not located on the top floor. Then $P(A) = \frac{48}{56}$ and $P(B \mid A) = \frac{47}{55}$. Next, by the product rule, you find again the value 0.7325 for the probability that no top floor apartment is available:

$$P(A \text{ and } B) = P(A)P(B \mid A) = \frac{48}{56} \times \frac{47}{55} = 0.7325.$$

Example 2.7. Three boys and three girls are planning a dinner party. They agree that two of them will do the washing up, and they draw lots to determine which two it will be. What is the probability that two boys will wind up doing the washing up?

Solution. A useful solution strategy in probability is to see whether your problem is the same as another problem, for which the solution is more obvious. This is the situation, here. The sought-after probability is the same as the probability of getting two red balls when blindly choosing two balls from a bowl containing three red and three blue balls. If A represents the event that the first ball chosen is red, and B represents the event that the second ball chosen is red, then the sought-after probability is equal to $P(A \text{ and } B)$. Thus, using the basic formula $P(A \text{ and } B) = P(A)P(B \mid A)$, you find that $P(\text{two boys will do the washing up}) = \frac{3}{6} \times \frac{2}{5} = \frac{1}{5}$.

An obvious extension of the product formula is

$$P(A_1 \text{ and } A_2 \text{ and } \ldots \text{ and } A_n)$$
$$= P(A_1) \times P(A_2 \mid A_1) \times \cdots \times P(A_n \mid A_1 \text{ and } A_2 \text{ and} \ldots \text{ and } A_{n-1}).$$

This useful extension is illustrated with the following example.

Example 2.8. What is the probability that you must pick five or more cards from a shuffled deck of 52 cards before getting an ace?

Solution. Noting that the sought-after probability is nothing else than the probability of getting no ace among the first four picked cards, let A_i be the event that the ith picked card is not an ace for $i = 1, \ldots, 4$. The probability $P(A_1$ and A_2 and A_3 and $A_4)$ is the probability that five or more cards are needed to get an ace. This probability is calculated from the extended product formula with $n = 4$ and has the value $\frac{48}{52} \times \frac{47}{51} \times \frac{46}{50} \times \frac{45}{49} = 0.7187$.

An alternative calculation is as follows: let E_k be the event that the first $k - 1$ cards are non-aces and F_k be the event that the kth card is an ace. Then, the probability p_k of getting the first ace at the kth pick is $P(E_k$ and $F_k) = P(E_k)P(F_k \mid E_k)$. Verify yourselves

$$p_k = \frac{\binom{48}{k-1}}{\binom{52}{k-1}} \times \frac{4}{52 - (k - 1)} \quad \text{for } k = 1, 2, \ldots, 49,$$

which gives $\sum_{k=5}^{49} p_k = 0.7187$. It never hurts to solve a problem in different ways. It allows you to double check your answer.

The foregoing examples show that when you use an approach based on conditional probabilities to solve the problem, you usually go straight to work without first defining a sample space. The counting approach, however, does require the specification of a sample space. If both approaches are possible for a given problem, then the approach based on conditional probabilities will, in general, be simpler than the counting approach.

For events A and B with nonzero probabilities, the formula

$$\boxed{P(B \mid A) = \frac{P(A \text{ and } B)}{P(A)}}$$

quantifies how the original probability $P(B)$ changes when new information becomes available. If $P(B \mid A) = P(B)$, then the events A and B are said to be *independent*. An equivalent definition of independence is $P(A \text{ and } B) = P(A)P(B)$. The concept of independence will be further explored in Section 2.7.

Beginning students sometimes think that independent events A and B with nonzero probabilities are disjoint. This is not true. The explanation is that $P(A \text{ and } B) = 0$ if A and B are disjoint, whereas $P(A \text{ and } B) = P(A)P(B) > 0$ if A and B are independent.

Problem 2.10. Five friends are sitting at a table in a restaurant. Two of them order white wine, and the other three order red wine. The waiter has forgotten who ordered what and puts the drinks in random order before the five persons. What is the probability that each person gets the correct drink? (answer: $\frac{1}{10}$)

Problem 2.11. A bag contains 14 red cards and 7 black cards. You pick two cards at random from the bag. Verify that it is more likely to pick one red and one black card rather than two red cards. (answer: the probabilities are $\frac{14}{30}$ and $\frac{13}{30}$)

Problem 2.12. Someone has rolled two dice out of your sight. You ask this person to answer "yes or no" on the question whether there is a six among the two rolls. He truthfully answers, "yes." What is the probability that two sixes were rolled? (answer: $\frac{1}{11}$)

Problem 2.13. A prize is raffled among 10 people. In a pre-agreed order, each of them draws a lottery ticket from a bowl with 10 tickets, including one winning ticket. What is the probability that the kth person in the row will win the prize? (answer: $\frac{1}{10}$ for all k)

Problem 2.14. In a variation of the hilarious TV-show game of Egg Russian roulette, two participants are shown an egg box with four boiled eggs and two raw eggs in random order. They take turns taking an egg and smashing it upon their heads. What is the probability that the one who starts will be the first to smash a raw egg? (answer: 0.6) How does this probability change when there are five boiled eggs and one raw egg? (answer: 0.5)

Problem 2.15. Four British teams are among the eight teams that have reached the quarter-finals of the Champions League soccer. What is the probability that the four British teams will avoid each other in the quarter-finals draw if the eight teams are paired randomly? (answer: $\frac{8}{35}$) *Hint*: think of a bowl containing four red and four blue balls where you remove each time two randomly chosen balls from the bowl. What is the probability that you remove each time a red and a blue ball? Solving a probability problem becomes often simpler by casting the problem into an equivalent form. Analogical thinking is creative thinking!

Problem 2.16. If you pick at random two children from the Johnson family, the chances are 50% that both children have blue eyes. How many children does the Johnson family have, and how many of them have blue eyes? (answer: 4 and 3)

Problem 2.17. Your friend shakes thoroughly two dice in a dice-box. He then looks into the dice-box. Your friend is honest and always tells you if he sees a six, in which case he bets with even odds that both dice show an even number. Is the game favorable to you? (answer: yes, your probability of winning is $\frac{6}{11}$)

2.3 The law of conditional probability

Suppose that a closed box contains one ball. This ball is white. An extra ball is added to the box, and the added ball is white or red with equal chances. Next one ball is blindly removed from the box. What is the probability that the removed ball is white? A natural reasoning is as follows. The probability of removing a white ball is 1 if a white ball is added to the box and is $\frac{1}{2}$ if a red ball is added to the box. It is intuitively reasonable to average these conditional probabilities over the probability that a white ball is added and the probability that a red ball is added. The latter two probabilities are both equal to $\frac{1}{2}$. Therefore,

$$P(\text{the removed ball is white}) = 1 \times \frac{1}{2} + \frac{1}{2} \times \frac{1}{2} = \frac{3}{4}.$$

This is an application of the law of conditional probability. This law calculates a probability $P(A)$ with the help of appropriately chosen conditioning events B_1 and B_2. These events must be such that event

A can occur only if one of events B_1 and B_2 has occurred, and events B_1 and B_2 must be disjoint. Then, $P(A)$ can be calculated as

$$P(A) = P(A \mid B_1)P(B_1) + P(A \mid B_2)P(B_2).$$

This is the *law of conditional probability*. It is a very useful rule to calculate probabilities.[9] The extension of the rule to more than two conditioning events B_i is obvious. In general, the choice of the conditioning events is self-evident. In the above example of the two balls, the conditioning event B_1 is the event that a white ball was added to the box and B_2 is the event that a red ball was added.

Example 2.9. A drunkard removes two randomly chosen letters of the message HAPPY HOUR that is attached on a billboard outside a pub. His drunk friend puts the two letters back in a random order. What are the chances of HAPPY HOUR appearing again?

Solution. Let A be the event that the message HAPPY HOUR appears again. In order to calculate $P(A)$, it is obvious to condition on the two events B_1 and B_2, where B_1 is the event that two identical letters were removed and B_2 is the event that two different letters were removed. In order to apply the law of conditional probability, you need to know the probabilities $P(B_1)$, $P(B_2)$, $P(A \mid B_1)$ and $P(A \mid B_2)$. The latter two probabilities are easy: $P(A \mid B_1) = 1$ and $P(A \mid B_2) = \frac{1}{2}$. The probabilities $P(B_1)$ and $P(B_2)$ require some more thought. It suffices to determine $P(B_1)$ because $P(B_2) = 1 - P(B_1)$. The probability $P(B_1)$ is the sum of the probability that the drunkard has removed the two H's, and the probability that the drunkard has removed the P's. Each of the latter two probabilities is equal to $\frac{2}{9} \times \frac{1}{8} = \frac{1}{36}$, by the product rule. Thus,

$$P(B_1) = \frac{1}{18} \text{ and } P(B_2) = \frac{17}{18}.$$

Then, by the law of conditional probability,

$$P(A) = 1 \times \frac{1}{18} + \frac{1}{2} \times \frac{17}{18} = \frac{19}{36},$$

---

[9]The proof is simple. Since A can only occur if one of events B_1 or B_2 has occurred and events B_1 and B_2 are disjoint, $P(A) = P(A \text{ and } B_1) + P(A \text{ and } B_2)$ (by Axiom 3 in Section 2.1). The product rule next leads to $P(A) = P(A \mid B_1)P(B_1) + P(A \mid B_2)P(B_2)$.

and so the probability that the message appears again is $\frac{19}{36}$.

The following problems ask you to apply the law of conditional probability.

Problem 2.18. Michael arrives home on time with probability 0.8. If Michael does not arrive home on time, the probability that his dinner is burnt is 0.5; otherwise, this probability is 0.15. What is the probability that Michael's dinner will be burnt? (answer: 0.22)

Problem 2.19. Your friend has chosen at random a card from a standard deck of 52 cards, but keeps this card concealed. You have to guess which of the 52 cards it is. Before doing so, you can ask your friend the question whether the chosen card is red or the question whether the card is the ace of spades. Your friend will answer truthfully. What question would you ask? (answer: the probability of a correct guess is $\frac{1}{26}$ in both cases)

Problem 2.20. A toss of a biased coin gives heads with probability p and tails with probability $q = 1 - p$. What is the probability that more than n tosses are needed to get both heads and tails? (answer: $p^n + q^n$)

Problem 2.21. You are the first to spin a wheel of fortune once or twice and then followed by your friend. The wheel has 20 sections numbered as $5, 10, \ldots, 100$. Each number has an equal chance of being the outcome of a spin. The winner is the one with a total score closest to 100 without exceeding it. In the event of a tie, you will be declared the winner. What is your maximum chance of winning? (answer: your chance of winning is 0.4824 if you do a second spin if the first spin scores less than 55 points and otherwise.)

Problem 2.22. Arthur and Mark play a series of games until one of the players has won two games more than the other player. Any game will be won by Arthur with probability p and by Mark with probability $q = 1 - p$. The outcomes of the games are independent of each other. What is the probability that Arthur will be the winner of the match? (answer: $p^2/(1 - 2pq)$)

2.4 Bayesian approach to inference

The Bayesian view of probability is interwoven with conditional probability. Bayes' formula, which is nothing else than logical thinking, is the most important rule in Bayesian probability.[10] To introduce this rule, consider again Problem 2.18. In this problem, Michael finds his dinner burnt (event A) with probability 0.5 if he does not arrive home on time (event B). That is, $P(A \mid B) = 0.5$. Suppose you are asked to give the probability $P(B \mid A)$, being the conditional probability that Michael did not arrive home on time given that his dinner is burnt. In fact you are asked to reason back from effect to cause. Then, you are in the area of Bayesian probability. The *basic form of Bayes' rule* is

$$P(B \mid A) = \frac{P(B)P(A \mid B)}{P(A)}$$

for any two events A and B with $P(A) > 0$. The derivation of this formula is strikingly simple. The basic form of Bayes' rule follows directly from the definition of conditional probability:

$$P(B \mid A) = \frac{P(B \text{ and } A)}{P(A)} = \frac{P(B)P(A \mid B)}{P(A)}.$$

In Problem 2.18, the values of $P(B)$ and $P(A \mid B)$ were given as 0.2 and 0.5, and the value of $P(A)$ can be calculated as 0.22. Thus the probability that Michael did not arrive home on time given that his dinner is burnt is equal to

$$P(B \mid A) = \frac{0.2 \times 0.5}{0.22} = \frac{5}{11}.$$

You see that Bayes' rule enables you to reason back from effect to cause in terms of probabilities. Many interesting queries are matters

---

[10]This formula is named after the English clergyman Thomas Bayes (1702–1762) who derived a special case of the formula. The formula in its general form was first written down by Pierre Simon Laplace (1749–1827). The famous British scientist Sir Harold Jeffreys (1891–1989) once stated that Bayes' formula is to the theory of probability what the Pythagorean theorem is to geometry.

of statistical inference, where the aim is to reason "backwards" from observed effects to unknown causes. In medical diagnosis, for example, the physician records a set of symptoms and must identify the underlying disease. Bayes' rule is the answer to such questions.

There are various versions for Bayes' rule. The most insightful version is the Bayes' rule in odds form. This version is mostly used in practice. Before stating Bayes' rule in odds form, the concept of odds will be discussed. Let G be any event that will occur with probability p, and so event G will not occur with probability $1 - p$. Then the *odds* of event G are defined by:

$$o(G) = \frac{p}{1-p}.$$

Conversely, the odds $o(G)$ of an event G determines $p = P(G)$ as

$$p = \frac{o(G)}{1 + o(G)}.$$

For example, an event G with probability $\frac{2}{3}$ has odds 2 (it is often said the odds are 2:1 in favor of event G), while an event with odds $\frac{2}{9}$ (odds are 2:9) has a probability $\frac{2}{11}$ of occurring.

Bayes' rule in odds form will be formulated in terms of events H (hypothesis) and E (evidence) rather than events A and B. Also, the standard notation $\overline{H}$ is used for the event that event H does not occur. Then, *Bayes' rule in odds form* reads as[11]

$$\frac{P(H \mid E)}{P(\overline{H} \mid E)} = \frac{P(H)}{P(\overline{H})} \times \frac{P(E \mid H)}{P(E \mid \overline{H})}.$$

What does this formula say and how to use it? This is easiest explained with the help of an example. Suppose that a team of divers

---

[11]This rule is obtained as follows. The basic form of the formula of Bayes gives that $P(H \mid E) = P(H)P(E \mid H)/P(E)$ and $P(\overline{H} \mid E) = P(\overline{H})P(E \mid \overline{H})/P(E)$. Taking the ratio of these two expressions, $P(E)$ cancels out and you get Bayes' rule in odds form. The derivation shows that the formula is also true when $\overline{H}$ would not be the complement of H. That is, for any two hypotheses H_1 and H_2, the general Bayes formula $\frac{P(H_1 \mid E)}{P(H_2 \mid E)} = \frac{P(H_1)}{P(H_2)} \times \frac{P(E \mid H_1)}{P(E \mid H_2)}$ applies.

believes that a sought-after wreck will be in a certain sea area with a probability of $p = 0.4$. A search in that area will detect the wreck with a probability of $d = 0.9$ if it is there. What is the revised probability of the wreck being in the area when the area is searched and no wreck is found? To answer this question, let hypothesis H be the event that the wreck is in the area in question and thus $\overline{H}$ is the event that the wreck is not in that area. Before the search takes place, your belief is that the events H and $\overline{H}$ have probabilities $P(H) = 0.4$ and $P(\overline{H}) = 0.6$. These probabilities are called *prior probabilities*. The ratio

$$\frac{P(H)}{P(\overline{H})} = \frac{0.4}{0.6}$$

is the *prior odds* of hypothesis H. These odds will change if additional information becomes available. Denote by evidence E the event that the search for the wreck is not successful. The probabilities $P(E \mid H)$ and $P(E \mid \overline{H})$ are given by $1 - 0.9$ and 1. The ratio

$$\frac{P(E \mid H)}{P(E \mid \overline{H})} = \frac{0.1}{1}$$

is called the *likelihood ratio* or *Bayes factor*. In the example, it has a value less than 1 and so the evidence does not support the hypothesis H. The *posterior odds* of hypothesis H can now be calculated as

$$\frac{P(H \mid E)}{P(\overline{H} \mid E)} = \frac{0.4}{0.6} \times \frac{0.1}{1} = \frac{1}{15}.$$

Then, by $P(\overline{H} \mid E) = 1 - P(H \mid E)$, you get that the *posterior probability* of hypothesis H is

$$P(H \mid E) = \frac{1/15}{1 + 1/15} = \frac{1}{16}.$$

This is the revised value of the probability that the wreck is in the area in question after the futile search. In general, the posterior probability $P(H \mid E)$ gives the updated value of the probability that hypothesis H is true after that additional information has become available through the evidence event E. Bayesian updating — revising an estimate when new information is available — is a key concept in statistics and data science.

Example 2.10. An athlete selected by lot has to go to the doping control. On average, 7 out of 100 athletes use doping. The doping test gives a positive result with a probability of 96% if the athlete has used doping and with a probability of 5% if the athlete has not used doping. Suppose that the athlete gets a negative test result. What is the probability that the athlete has nevertheless used doping?

Solution. Let the hypothesis H be the event that the athlete has used doping. The prior probabilities are $P(H) = 0.07$ and $P(\overline{H}) = 0.93$. Let E be the event that the athlete has a negative test result. Then, $P(E \mid H) = 0.04$ and $P(E \mid \overline{H}) = 0.95$. The posterior odds of the hypothesis H are

$$\frac{P(H \mid E)}{P(\overline{H} \mid E)} = \frac{0.07}{0.93} \times \frac{0.04}{0.95} = 0.003169.$$

Thus, the revised value of the probability of doping use notwithstanding a negative test result is

$$P(H \mid E) = \frac{0.003169}{1 + 0.003169} = 0.003159,$$

or rather about 0.32%. A very small probability indeed.

The posterior probability of 0.32% can also be calculated without using conditional probabilities and Bayes' rule. The alternative calculation is based on the method of *expected frequencies*. This method is also easy to understand by the layman. Imagine a very large number of athletes that are selected by lot for the doping control, say 10 000 athletes. On average, 700 of these athletes have used doping, and 9 300 athletes have not used doping. Of these 700 athletes, $700 \times 0.04 = 28$ athletes test negative on average, whereas $9\,300 \times 0.95 = 8\,835$ athletes of the other 9 300 athletes test negative on average. Thus, a total of $28 + 8\,835 = 8\,863$ athletes test negative and among those 8 863 athletes, there are 28 doping users. Therefore, the probability that an athlete has used doping notwithstanding a negative test result is $\frac{28}{8\,863} = 0.003159$. The same probability as found with Bayes' rule. A similar reasoning shows that the probability that an athlete with a positive test result has not used doping is $\frac{465}{465+672} = 0.40987$ (verify!).

The following example is a fun application of Bayes' rule.

Example 2.11. Two closed boxes are placed in front of you. One box contains nine $1 bills and one $5 bill, while the other box contains two $1 bills and one $100 bill. You choose at random one box. Then, out of your sight, two bills are randomly picked out of the chosen box. It appears that these two bills are $1 bills. Next, you get the opportunity to pick a third bill out of one of the two boxes. Should you stick to the chosen box, or should you switch to the other box when you want to maximize the probability of picking the $100 bill?

Solution. Let the hypothesis H be the event that you have chosen the box with the $100 bill and $\overline{H}$ be the event that you have not chosen the $100 bill box. The prior probabilities are $P(H) = P(\overline{H}) = \frac{1}{2}$. Let the evidence E be the event that the two bills taken out of the chosen box are $1 bills. Then $P(E \mid H) = \frac{2}{3} \times \frac{1}{2} = \frac{1}{3}$ and $P(E \mid \overline{H}) = \frac{9}{10} \times \frac{8}{9} = \frac{4}{5}$. Thus, by Bayes' rule in odds form,

$$\frac{P(H \mid E)}{P(\overline{H} \mid E)} = \frac{1/2}{1/2} \times \frac{13}{45} = \frac{5}{12}.$$

This leads to $P(H \mid E) = \frac{5/12}{1+5/12} = \frac{5}{17}$ and $P(\overline{H} \mid E) - 1 - \frac{5}{17} = \frac{12}{17}$. Thus, $P(H \mid E) \times 1 = \frac{5}{17}$ is the probability of picking the $100 bill as third bill if you stick to the chosen box. If you switch to the other box, your win probability is $P(\overline{H} \mid E) \times \frac{1}{3} = \frac{4}{17}$. Therefore you better stick to the chosen box. A surprising finding! Can you explain it?

The following problems ask you to apply Bayes' rule in odds form. You should first identify the hypothesis H and the evidence E.

Problem 2.23. An oil explorer performs a seismic test to determine whether oil is likely to be found in a certain area. The probability that the test indicates the presence of oil is 90% if oil is indeed present in the test area, while the probability of a false positive is 15% if no oil is present in the test area. Before the test is done, the explorer believes that the probability of presence of oil in the test area is 40%. What is the revised probability of oil being present in the test area given that the test is positive? (answer: 0.8)

Problem 2.24. In a certain region, it rains on average once in every ten days during the summer. Rain is predicted on average for 85% of the days when rainfall actually occurs, while rain is predicted on average for 25% of the days when it does not rain. The weather on one day does not depend on the weather on previous days. Rain is predicted for tomorrow. What is the probability of rainfall actually occurring on that day? (answer: $\frac{17}{62}$)

Problem 2.25 A murder is committed. The perpetrator is either one or the other of the two persons X and Y. Both persons are on the run from authorities, and after an initial investigation, both fugitives appear equally likely to be the perpetrator. Further investigation reveals that the actual perpetrator has blood type A. Ten percent of the population belongs to the group having this blood type. Additional inquiry reveals that person X has blood type A, but offers no information concerning the blood type of person Y. In light of this new information, what is the probability that person X is the perpetrator? (answer: $\frac{1}{11}$)

Problem 2.26. One fish is contained in an opaque fishbowl. The fish is equally likely to be a piranha or a goldfish. A sushi lover throws a piranha into the fishbowl alongside the other fish. Then, immediately, before either fish can devour the other, one of the fish is blindly removed from the fishbowl. The removed fish appears to be a piranha. What is the probability that the fish that was originally in the bowl by itself was a piranha? (answer: $\frac{2}{3}$)

Problem 2.27. On the island of liars, each inhabitant lies with probability $\frac{2}{3}$. You overhear an inhabitant making a statement. Next, you ask another inhabitant whether the inhabitant you overheard spoke truthfully. What is the probability that the inhabitant you overheard indeed spoke truthfully given that the other inhabitant says so? (answer: $\frac{1}{5}$)

Problem 2.28. You have two symmetric dice in your pocket. One die is a standard die, and the other die has each of the three numbers 2, 4, and 6 twice on its faces. You randomly pick one die from your pocket without looking. Someone else rolls this die and informs you that a 6 has shown up. What is the revised value of the probability that you have picked the standard die? How does this probability change if the die is rolled a second time and a 6 appears again? (answer: $\frac{1}{3}$ and $\frac{1}{5}$)

Problem 2.29. Someone visits the doctor because he fears having a very rare disease. This disease occurs in only 0.1% of the population. The doctor proposes a test that correctly identifies 99% of the people who have the disease, and only incorrectly identifies 1% of the people who don't have the disease. Suppose the person in question has tested positive. What is the probability that he actually has the disease? (answer: 9.02%) How does the answer change if a second independent test is also positive? (answer: 90.75%)

2.4.1 Real-life cases of Bayesian thinking

This subsection presents cases from law and medicine in which conditional probabilities are incorrectly used. Attention will be paid to the prosecutor's fallacy often associated with miscarriages of justice.

The case of Sally Clark: a miscarriage of justice

Sally Clark was arrested in 1999 after her second child, who was a few months old, died, ostensibly by cot death, just as her first child had died a year earlier. She was accused of suffocating both children. During the trial the prosecutor called a famous pediatrician as an expert. He stated that the chance of cot death of a child was 1 in 8 543, and that the chance of two cot deaths in the same family was $\left(\frac{1}{8543}\right)^2$ or about 1 in 73 million. The prosecutor argued that, beyond any reasonable doubt, Sally Clark was guilty of murdering her two children, and the jury sentenced her to life imprisonment, though there was no other evidence that Sally Clark had killed her two children. This is a classic example of the 'prosecutor's fallacy'. The probability of innocence given the death of the two children — the probability that matters — is confused with the tiny probability that in the same family two infant children will die of sudden infant death syndrome.

The conviction of Sally Clark led to great controversy, and several leading British statisticians threw themselves into the case. The statisticians came up with various estimates for Sally Clark's chance of innocence, and all these estimates showed that the condemnation of her was not beyond reasonable doubt. The formula of Bayes was the basis of the calculations of the statisticians. How did this work? Let H be the event that Sally Clark is guilty and the evidence E be the event that both of her children died in the first few months of their lives. The probability that matters is the conditional probability $P(H \mid E)$. To get this probability, you need prior probabilities $P(H)$ and $P(\overline{H})$ together with likelihood ratio $P(E \mid H)/P(E \mid \overline{H})$. The assumption is made that murder by the mother (hypothesis H) and cot death (hypothesis $\overline{H}$) are the only two possibilities for the death of the two children. Of course, $P(E \mid H) = 1$. The pediatrician called as expert gave the estimate $\frac{1}{8\,543} \times \frac{1}{8\,543}$ for $P(E \mid \overline{H})$, but this estimate assumes independence between both deaths. However, a cot death in a family increases the likelihood that a subsequent birth in the family will also die of cot death. In an article in the British Medical Journal, it was made plausible that a factor of 5 applies to the increased chance. Thus, the probability $P(E \mid \overline{H})$ is estimated by

$$P(E \mid \overline{H}) = \frac{1}{8\,543} \times \frac{5}{8\,543} \approx 6.85 \times 10^{-8},$$

or, about 1 in 14.8 million, which is still a very small probability. However, this probability should be weighed with the very small prior probability that a mother will kill both of her children at the beginning of their first year of life by suffocation. How do you get a good estimate for the prior probability $P(H)$? This is not simple. However, on the basis of statistical data, an upper bound for the prior probability $P(H)$ can be estimated. Instead of asking how often mothers in a family like the Clarks kill their first two children in their first year of life, the question can be answered on how often mothers kill one or more of their children of any age. Data available in the U.S. Statistics give about 100 cases per year in the U.S. In the U.S. there are about 120 million adult women, and about half of them have children, so about 1 in 0.6 million American women murders one or more of their children. The frequency of murders in America is about 4 times as large as in England. This leads to the estimate that about 1 in 2.4 million women in England kills one or more of their children. This is, of course, an overestimate of the prior probability $P(H)$. If you nevertheless take $P(H) = \frac{1}{2.4 \times 10^6}$, then you find

$$\frac{P(H \mid E)}{1 - P(H \mid E)} \approx \frac{1/(2.4 \times 10^{-6})}{1 - 1/(2.4 \times 10^{-6})} \times \frac{1}{6.85 \times 10^{-8}} \approx 6.08.$$

Thus, the posterior probability of Sally Clark's guilt is given by

$$P(H \mid E) \approx \frac{6.08}{1 + 6.08} = 0.859.$$

In other words, the posterior probability of Sally Clark's innocence is estimated by 0.141. In fact, this is an underestimate, since 0.859 is an overestimate of the probability of Sally Clark's guilt.[12] So a probability of 14.1% or more is a reasonable estimate for the probability of Sally Clark's innocence. This is of course no base for a conviction when there is no other evidence. Despite the arguments that statisticians presented, Sally Clark lost the appeal against her conviction. But in 2003 she was acquitted after it came out that her second child had a bacterial infection in the brain at the time of his death, a fact that was withheld from the defense in the earlier trial. The tragic event surrounding Sally Clark is similar with the miscarriage of justice that took place in the Netherlands around the nurse Lucia de Berk who was wrongly accused of murdering a number of her patients who died during her night shifts.

People v. Collins

An older famous example of the prosecutor's fallacy is People v. Collins, an American robbery trial. On June 18, 1964, Juanita Brooks was attacked in an alley near to her home in Los Angeles and her purse stolen. A witness reported that a white woman running from the scene was blond, had a pony tail, and fled from the scene in a yellow car driven by a black man with a beard and a mustache. Police arrested a couple, Janet and Mark Collins, which fit the description. Unfortunately for the prosecutor, neither Juanita Brooks

---

[12] An overestimate of the prior $P(H)$ always results in an overestimate of the posterior $P(H \mid E)$. This intuitive result can be deduced from Bayes' formula, using the fact that $a/(1-a) < b/(1-b)$ if and only if $a < b$ when $0 < a, b < 1$.

nor the witness could make a positive identification of either of the defendants. At the trial, following testimony by a college mathematics instructor, the prosecutor provided the following probabilities of occurrence of the reported characteristics: girl with blond hair $\frac{1}{3}$, girl with ponytail $\frac{1}{10}$, yellow car $\frac{1}{10}$, man with mustache $\frac{1}{4}$, black man with beard $\frac{1}{10}$, and interracial couple in car $\frac{1}{1000}$. Although the instructor had told the prosecutor that these individual probabilities could not simply be multiplied by each other, the prosecutor multiplied them to arrive at a joint probability of

$$\frac{1}{3} \times \frac{1}{10} \times \frac{1}{10} \times \frac{1}{4} \times \frac{1}{10} \times \frac{1}{1\,000} = \frac{1}{12\,000\,000}.$$

In his summation, the prosecutor emphasized the extreme unlikelihood that a couple other than the defendants had all these characteristics. Impressed by the long odds, the jury convicted the couple for second-degree robbery. But did they make the right decision? The answer is no! The fundamental flaw in the prosecutor's reasoning was to equate the probability of innocence of the Collins couple with the probability that a randomly chosen couple would have the six characteristics in question. Moreover, the individual probabilities of the characteristics were unfounded, and by dependence between the characteristics (beards and mustaches are not independent events), it was wrong to multiply them to arrive at a probability of 1 in 12 million. Defense appealed and the California Supreme Court reversed the conviction, very critical of the statistical reasoning used and the way it was put to the jury.

O. J. Simpson trial

A remarkable example of confusing conditional probabilities happened in the O. J. Simpson trial. This trial, regarded by many as 'the trial of the century', dominated the news for more than a year and was broadcast on television. In 1994, O. J. Simpson, an actor and former American football star, was accused of murdering his ex-wife, Nicole Brown. The trial started with prosecution proving that O. J. Simpson has a history of physical violence against his ex-wife Nicole. The famous lawyer Alan Dershowitz countered for the defense. Dershowitz argued, "Only about one in 2500 men who batter their domestic partners go on to murder them, so the fact that O. J. Simpson battered his wife is irrelevant to the case." This was a clever trick to fool the non-mathematical jury. The defense lawyer confused the jury with asking the question: what is the probability of a man murdering his partner given that he previously battered her? This conditional probability is indeed about 0.04%, but it is not the right statistic. It ignores the crucial fact that Nicole Brown was actually murdered. The real question is: what is the probability that a man murdered his wife given that he previously battered her and she was murdered? That conditional probability P(abusive husband is guilty | wife is murdered) turns out to be very far from 0.04%. It can be approximated by Bayes' rule. Instead, let us use the expected frequency approach. Imagine a sample of 100 000 battered women. According to Dershowitz's number of 1 in 2500, you can expect about 40 of these women to be murdered by their violent partners in a given year. Let us take as rough estimate that an additional 5 of the 100 000 battered women will be killed by someone else, on the basis of the fact that the murder rate for *all* women in the United States at the time of the trial was about 1 in 20 000 per year. Then the conditional probability that a man murdered his wife given that he previously battered her and she was murdered can be approximated by $\frac{40}{40+5} = \frac{8}{9}$, or about 89%. This probability is much and much larger than the probability of 0.04% in the defense lawyer's argument. The estimated probability of 89% shows that the fact that O. J. Simpson had physically abused his wife in the past was certainly very relevant to the case. This probability should not be confused with the probability that O. J. Simpson did it. Many other factors also determine this probability.

2.4.2 Bayesian statistics vs. classical statistics

Bayesian probability is the basis of Bayesian statistics. This field of statistics is very different from the field of classical statistics dealing with hypothesis-testing. Imagine that a new medication is being tested on a given number of patients, and that it appears to be effective for a number of them. You want to know if this means that the medication works. In classical statistics, you would start with the assumption that mere fluke is the cause of the test results (this is called the null hypothesis). The null hypothesis is then tested using the so-called p-value, being the probability of getting data that are *at least* as good as the observed data if the null hypothesis would be true. If the p-value is below some threshold value — the value 0.05 is often used as a cut-off value — the null hypothesis is rejected and it is assumed that the medication is effective (it is said that the findings are 'statistically significant'). The p-value, however, does not tell you what the probability is that the new medication is not effective. And this is, in fact, the probability you really want to know. Scientific studies have shown that the p-value can give a highly distorted picture of this probability. The probability that the medication is not effective can be considerably larger than the p-value. As a consequence, you must be careful with drawing a conclusion when the p-value is just below 0.05; the test $p < 0.05$ is not a litmus test. As such the test was never intended, but it was meant as a signal to investigate matters further. In the case that the p-value is extremely small and the study is carefully designed, then further investigations are not necessary. Physicists reported the 'discovery' of the Higgs boson in 2012 after a statistical analysis of the data showed that they had attained a confidence level of about a one-in-3.5 million probability that the experimental results would have been obtained if there were no Higgs particle. Again, in all clarity, this p-value is not the probability that the Higgs boson doesn't exist. Another extremely small p-value occurred in the famous 1954 Salk vaccine polio study. In this statistical study, a double-blind study was conducted with two groups of 200 000 children each. This huge number of children was required to get enough observations of polio for statistical analysis. A total of 142 children in the placebo group developed polio and 57 children in the vaccine group. The p-value was on the order of 10^{-9} and removed any doubts about the vaccine's efficacy.

Bayesian statistics enables you to give a judgment about the probability that the medication works. The judgment uses the generic Bayesian formula:

$$p(\theta \mid data) = \frac{p(data \mid \theta)p(\theta)}{p(data)}.$$

What is the meaning of the elements of this formula? This formula determines the posterior probability distribution $p(\theta \mid data)$ of an unknown parameter θ. For example, you can think of θ as the percentage of people for whom a new medication is working. To estimate the posterior distribution, you need data from test experiments. Before the tests are done, you should specify a prior probability distribution $p(\theta)$ on the parameter θ. The (subjective) prior probabilities represent the uncertainty in your knowledge about the true value of θ. It is your knowledge about the parameter that is modeled as random, not the parameter itself. The use of priors distinguishes Bayesian statistics from classical statistics. The so-called likelihood $p(data \mid \theta)$ is the probability of finding the observed data for a given value of θ, and $p(data)$ is obtained by averaging $p(data \mid \theta)$ over the prior probabilities $p(\theta)$. This describes in general terms how Bayesian statistics works. To illustrate this, consider the following experiment. Suppose that there is a reason to believe that a coin might be slightly biased towards heads. To test this, you decide to throw the coin 1000 times. Before performing the experiment, you express your uncertainty about the unbiasedness of the coin by assuming that the probability of getting heads in a single toss of the coin can take on the values $\theta = 0.50, 0.51$, and 0.52 with prior probabilities $p(\theta) = \frac{1}{2}, \frac{1}{3}$, and $\frac{1}{6}$, respectively. Next, the experiment is performed and 541 heads are obtained in 1000 tosses of the coin. The likelihood of getting 541 heads in 1000 tosses is

$$p(\text{data} \mid \theta) = \binom{1\,000}{541} \theta^{541}(1-\theta)^{459} \quad \text{for } \theta = 0.50, 0.51, 0.52.$$

By the law of conditional probability,

$$p(\text{data}) = p(\text{data} \mid 0.50)p(0.50) + p(\text{data} \mid 0.51)p(0.51)$$
$$+ p(\text{data} \mid 0.52)p(0.52).$$

The Bayes formula $p(\theta \mid \text{data}) = p(\text{data} \mid \theta)p(\theta)/p(\text{data})$ now gives that the posterior probability of a fair coin is

$$p(0.50 \mid \text{data}) = 0.1282.$$

That is, your posterior belief that the coin is fair equals 0.1282. In classical statistics, one would compute the probability of getting 541 or more heads in 1000 tosses of the coin under the hypothesis that the coin is fair. This excess probability is equal to 0.0052. Many classical statisticians would consider this small p-value as significant evidence that the coin is biased towards heads. However, your subjective Bayesian probability of 0.1282 for the hypothesis of a fair coin is not strong enough evidence for such a conclusion. The difference in the conclusions can be explained as follows. The p-value is based on the set of all possible observations that cast as much or more doubt on the hypothesis than the actual observations do. It is not possible to base the p-value only on the actual data because it frequently happens that all individual outcomes have such small probabilities that every outcome would look significant. The inclusion of unobserved data means that the resulting p-value may greatly exaggerate the strength of evidence against the hypothesis.

The Bayesian approach is used more and more in practice. Nowadays, Bayesian methods are used widely to address pressing questions in diverse application areas such as astrophysics, actuarial sciences, neurobiology, weather forecasting, spam filtering, and criminal justice. The next subsection discusses a practically useful Bayesian heuristic that is often used in data science.

2.4.3 Naive Bayes in data analysis

Naive Bayes is a simple machine learning method that is used to classify objects based on features. For example, an email can be classified as spam/non-spam by the words in it. The model is called naive because it naively assumes conditional independence of the features given the class of the target variable, regardless of possible correlations between the features. The target variable is often a binary variable with yes/no values. Despite the fact that the independence assumption is often inaccurate in reality, naive Bayes is remarkably useful in practice.

Let us explain naive Bayes by using the application area of spam filtering. Particular words have high probabilities of occurring in spam email. For instance, most email users will frequently encounter words such as viagra, replica, sex, etc. in spam email, but will seldom see them in other email. The filter doesn't know these probabilities in advance, and must first be trained so that it can be built upon. Suppose that for each word W in a set of relevant words, the following probabilities have been estimated using sufficiently large, representative data sets of spam and non-spam messages:

- $P(W \mid \text{spam})$, the probability that a spam message will contain the word W.

- $P(W \mid \text{no spam})$, the probability that a non-spam message contain the word W.

- $P(\text{spam})$, the prior probability that a message is spam.

- $P(\text{no spam})$, the prior probability that a message is non-spam.

How to combine the individual word probabilities to estimate the probability that an email with a particular set of words is spam or non-spam? Suppose an e-mail contains the words $w_1, \ldots, w_n$, which can appear in both spam and non-spam emails. To determine whether the email is spam or not, you need $P(\text{spam} \mid w_1, \ldots, w_n)$ and $P(\text{no spam} \mid w_1, \ldots, w_n) = 1 - P(\text{spam} \mid w_1, \ldots, w_n)$, where

$$P(\text{spam} \mid w_1, \ldots, w_n) = \text{the probability that the email is spam}$$
$$\text{given that it contains the words } w_1 \ldots, w_n.$$

By Bayes' rule in odds form,

$$\frac{P(\text{spam} \mid w_1, \ldots, w_n)}{P(\text{no spam} \mid w_1, \ldots, w_n)} = \frac{P(\text{spam})}{P(\text{no spam})} \times \frac{P(w_1, \ldots, w_n \mid \text{spam})}{P(w_1, \ldots, w_n \mid \text{no spam})}.$$

Naive Bayes now comes into the picture. Assuming that the words are conditionally independent given the target variable spam/no spam, this Bayes' formula gets the computationally tractable form:

$$\frac{P(\text{spam} \mid w_1, \ldots, w_n)}{P(\text{no spam} \mid w_1, \ldots, w_n)}$$
$$= \frac{P(\text{spam})}{P(\text{no spam})} \times \frac{P(w_1 \mid \text{spam}) \times \cdots \times P(w_n \mid \text{spam})}{P(w_1 \mid \text{no spam}) \times \cdots \times P(w_n \mid \text{no spam})}$$

The decision to classify the email as spam or non-spam depends on the ratio of $P(\text{spam} \mid w_1, \ldots, w_n)$ and $P(\text{no spam} \mid w_1, \ldots, w_n)$. A typical threshold value is 0.5.

2.5 The concept of random variable

In performing a chance experiment, one is often not interested in the particular outcome that occurs but in a specific numerical value associated with that outcome. Any function that assigns a real number to each outcome in the sample space is called a *random variable*.

The concept of random variable is always a difficult concept for the beginner. A random variable is not a variable in the traditional sense of the word, and actually it is a little misleading to call it a variable. Intuitively, a random variable is a function that takes on its value by chance. Formally, a random variable is defined as a function that assigns a numerical value to each element of the sample space. The observed value, or realization, of a random variable is completely determined by the realized outcome of the chance experiment, and consequently, probabilities can be assigned to the possible values of the random variable. A random variable gets its value only *after* the chance experiment has been done. *Before* the chance experiment is done, you can only speak of the probability that the random variable will take on a particular value. It is common to use uppercase letters such as X, Y, and Z to denote random variables, and lowercase letters x, y, and z to denote their numerical values.

Illustrative examples of random variables are:

• The number of goals to be scored in a soccer game.

• The number of major hurricanes that will hit the United States next year.

• The number of claims that will be submitted to an insurance company next month.

• The amount of rainfall that the city of London will receive next year.

• The time until radioactive material will emit a particle.

• The duration of your next mobile phone call.

The first three examples are examples of *discrete random variables* taking on a discrete number of values, and the other three examples describe *continuous random variables* taking on a continuum of values. In this book, the emphasis is on discrete random variables, but Chapter 3 also pays attention to continuous random variables.

Suppose that $a_1, a_2, \ldots, a_r$ are the possible values of a discrete random variable X. The notation $P(X = a_k)$ is used for the probability that the random variable X will take on the value a_k. These probabilities satisfy $\sum_{k=1}^{r} P(X = a_k) = 1$ and constitute the so-called *probability mass function* of the random variable X.

Example 2.12. What is the probability mass function of the random variable X denoting the sum of one roll of two fair dice?

Solution. It is helpful to think of a blue and a red die. The sample space of the chance experiment consists of the 36 outcomes (i, j) for $i, j = 1, \ldots, 6$, where i is the number rolled by the blue die and j is the number rolled by the red die. Each outcome is equally likely and gets assigned a probability of $\frac{1}{36}$. The random variable X takes on the value $i + j$ when the realized outcome is (i, j). The possible values of X are $2, \ldots, 12$. To find the probability $P(X = k)$, you must know the outcomes (i, j) for which the random variable X takes on the value k. For example, X takes on the value 7 for the outcomes $(1, 6)$, $(2, 5)$, $(3, 4)$, $(4, 3)$, $(5, 2)$, and $(6, 1)$. Each of these six outcomes has probability $\frac{1}{36}$. Thus, $P(X = 7) = \frac{6}{36}$. You are asked to verify

$$P(X = j) = \begin{cases} (j - 1)/36 & \text{for } 2 \leq j \leq 7, \\ (13 - j)/36 & \text{for } 8 \leq j \leq 12. \end{cases}$$

Can you explain why $P(X = 14 - j) = P(X = j)$ for $j = 2, \ldots, 6$?

Problem 2.30. Let the random variable X be the largest number rolled in one roll of two fair dice. What is the probability mass function of X? (answer: $P(X = j) = (2j - 1)/36$ for $j = 1, \ldots, 6$)

Problem 2.31. Each week the value of a particular stock either increases by 5% or decreases by 4% with equal chances, regardless of what happened before. The current value of the stock is \$100. What is the probability mass function of the value of the stock two weeks later? (answer: 0.25, 0.50, 0.25 for \$110.25, \$100.80, \$92.16)

2.6 Expected value and standard deviation

Suppose that $a_1, a_2, \ldots, a_r$ are the possible values of the discrete random variable X. The *expected value* (or *mean* or *average value*) of the random variable X is defined by

$$E(X) = a_1 \times P(X = a_1) + a_2 \times P(X = a_2) + \cdots + a_r \times P(X = a_r).$$

In words, $E(X)$ is a weighted average of the possible values that X can take on, where each value is weighted with the probability that X will take on that particular value.[13] The term 'expected value' can be misleading. This term should not be confused with the term 'most probable value'. Consider the following simple gambling game. A fair die is rolled. If a 6 appears, you get paid \$3; otherwise, you have to pay 60 cents. Define the random variable X as your gain (in dollars). Then X has the possible values 3 and -0.6 with probabilities $\frac{1}{6}$ and $\frac{5}{6}$, respectively. The most probable value of X is -0.6, but the expected value is

$$E(X) = 3 \times \frac{1}{6} - 0.6 \times \frac{5}{6} = 0.$$

As another example, let the random variable Y be the number of points in a single roll of a fair die. Then,

$$E(Y) = 1 \times \frac{1}{6} + 2 \times \frac{1}{6} + 3 \times \frac{1}{6} + 4 \times \frac{1}{6} + 5 \times \frac{1}{6} + 6 \times \frac{1}{6} = 3.5.$$

Using these two examples, the term 'average value' instead of 'expected value' can be explained. Suppose a fair die is repeatedly rolled. It will come as no surprise that the fraction of the number of rolls with outcome j goes to $\frac{1}{6}$ for all $j = 1, 2, \ldots, 6$ if the number of rolls gets larger and larger. Thus, if the number of rolls gets very large, the average number of points obtained per roll tends to $\frac{1}{6}(1 + 2 + \cdots + 6) = 3.5$, which is the expected value for a single roll. In the gambling game, your average gain per play tends to 0 if

---

[13]The idea of an expected value appears in the 1654 Pascal–Fermat correspondence, and this idea was elaborated on by the Dutch astronomer Christiaan Huygens (1625–1695) in his famous 1657 book *Ratiociniis de Ludo Aleae* (On Reasoning in Games of Chance).

the number of plays gets larger and larger. This is known as the *law of large numbers*. This law can be mathematically proved from the axioms and definitions of probability.

In many practical problems, it is helpful to interpret the expected value of a random variable as a long-term average. This is the case in the following example, which has its origin in World War II when a large number of soldiers had their blood tested for syphilis.

Example 2.13. A large number of individuals must undergo a blood test for a certain disease. The probability that a randomly selected person will have the disease is $p = 0.005$. In order to reduce costs, it is decided that the large group should be split into smaller groups, each made up of r persons, after which the blood samples of the r persons will be pooled and tested as one. The pooled blood samples will only test negative (disease free) if all of the individual blood samples were negative. If a test returns a positive result, then all of the r samples from that group will be retested, individually. What is the expected value of the number of tests that will have to be performed on one group of r individuals?

Solution. Define the random variable X as the number of tests that will have to be performed on a group of r individuals. The random variable X has the two possible values 1 and $r + 1$. The probability that X will take on the value 1 is equal to the probability that each individual blood sample will test negative, and this probability is $(1 - p) \times (1 - p) \times \cdots \times (1 - p) = (1 - p)^r$. This means that $P(X = 1) = 0.995^r$ and $P(X = r + 1) = 1 - 0.995^r$. Therefore,

$$E(X) = 1 \times 0.995^r + (r + 1) \times (1 - 0.995^r).$$

In other words, by pooling the blood samples of the r individuals, an average of $\frac{1}{r}\left(1 \times 0.995^r + (r + 1) \times \left(1 - 0.995^r\right)\right)$ tests per individual will be needed when many groups are tested. The average is minimal for $r = 15$ with 0.1391 as minimum value. Thus, the pooling of 15 individual blood samples saves about 86% on the number of tests necessary.

Example 2.14. You play the following game. A fair coin is tossed. If it lands heads, it will be tossed one more time; otherwise, it will be tossed two more times. You win eight dollars if heads does not come up at all, but you must pay one dollar each time heads does turn up. Is this a fair game?

Solution. The set $\{HH, HT, THH, THT, TTH, TTT\}$ is an obvious choice for the sample space, where H stands for heads and T for tails. The random variable X being defined as your net winnings in a game takes on the value 8 for outcome TTT, the value -1 for each of the outcomes HT, THT, and TTH, and the value -2 for each of the outcomes HH and THH. The probability $\frac{1}{2} \times \frac{1}{2} = \frac{1}{4}$ is assigned to each of the outcomes HH and HT, and the probability $\frac{1}{2} \times \frac{1}{2} \times \frac{1}{2} = \frac{1}{8}$ to each of the other four outcomes. Thus,

$$P(X = 8) = \frac{1}{8}, \; P(X = -1) = \frac{1}{4} + \frac{1}{8} + \frac{1}{8}, \text{ and } P(X = -2) = \frac{1}{4} + \frac{1}{8}.$$

This gives

$$E(X) = 8 \times \frac{1}{8} - 1 \times \frac{1}{2} - 2 \times \frac{3}{8} = -\frac{1}{4}.$$

The game is not fair. In the long run, you lose a quarter of a dollar per game.

Linearity property and the substitution rule

A very useful property of the expected value is the *linearity property*. This property states that

$$\boxed{E(aX + bY) = aE(X) + bE(Y)}$$

for any two random variables X and Y and any constants a and b, regardless whether or not there is dependence between X and Y. This result is valid for any type of random variables, as long as the expected values $E(X)$ and $E(Y)$ are well-defined. For the special case of discrete random variables each having only a finite number of possible values, a proof will be given in Section 2.10. More generally, for any random variables $X_1, \ldots, X_n$ and constants $c_1, \ldots, c_n$,

$$\boxed{E(c_1 X_1 + \cdots + c_n X_n) = c_1 E(X_1) + \cdots + c_n E(X_n).}$$

Another important result is the *substitution rule* (also called the *law of the unconscious statistician*), which will also be proved in Section 2.10. Suppose that $g(x)$ is a given function, e.g. $g(x) = x^2$ or $g(x) = \sin(x)$. Then $g(X)$ is also a random variable when X is a random variable. If X is a discrete random variable with possible values $a_1, \ldots, a_r$, then the substitution rule tells you that the expected value of $g(X)$ can be calculated through the intuitive formula:

$$E\big[g(X)\big] = \sum_{j=1}^{r} g(a_j) P(X = a_j).$$

Thus, you do not need to know the probability mass function of the random variable $g(X)$ in order to calculate its expected value. This is a very useful result. A special case of the substitution rule is

$$E(aX + b) = aE(X) + b \text{ for any constants } a \text{ and } b.$$

In many cases, the expected value of a random variable can be calculated by writing the random variable as a sum of *indicator random variables* that can take on only the values 0 and 1. This powerful trick is illustrated in the next example dealing with the celebrated *coupon collector's problem*.

Example 2.15. Each box of a newly launched breakfast cereal contains one baseball card, which is equally likely to be one of 50 different cards. What is the expected number of cards you are still missing after having purchased 100 boxes?

Solution. Define the random variable X as the number of cards you are still missing after having purchased 100 boxes. It is quite complicated to get $E(X)$ through the probability mass function of X. However, it is very easy to calculate $E(X)$ through the trick of writing X as $X = I_1 + I_2 + \cdots + I_{50}$, where $I_k = 1$ if the kth card is missing and $I_k = 0$ otherwise. By the linearity property of the expected value, you get

$$E(X) = \sum_{k=1}^{50} E(I_k).$$

The $E(I_k)$ are easy to get. For any k, the probability that a box does not contain the kth card is $\frac{49}{50}$ and so $P(I_k = 1) = \left(\frac{49}{50}\right)^{100}$. Since

$$E(I_k) = 1 \times P(I_k = 1) + 0 \times P(I_k = 0) = P(I_k = 1),$$

you have $E(I_k) = \left(\frac{49}{50}\right)^{100}$ for all k. Therefore, the expected value of the number of missing cards is $\sum_{k=1}^{50} E(I_k) = 50 \times \left(\frac{49}{50}\right)^{100} = 6.631$.

The coupon collector's problem from Example 2.15 is an application of the oft-used *balls-and-bins model*: each of $b = 100$ balls (purchases) is put randomly into one of $n = 50$ bins (baseball cards).

Variance and standard deviation

The expected value is an informative statistic in chance experiments that can be repeated indefinitely often. In other situations, it may be dangerous to rely merely on the expected value. Think of non-swimmer going into a lake that is, on average, 30 centimeters deep. A measure for the variability of a random variable X around its expected value $\mu = E(X)$ is the *variance* that is defined by

$$\boxed{\text{var}(X) = E\big[(X - \mu)^2\big].}$$

To illustrate, consider again random variable X with $P(X = 3) = \frac{1}{6}$ and $P(X = -0.6) = \frac{5}{6}$. The random variable X has the expected value $\mu = 0$ and its variance is equal to

$$E[(X - \mu)^2] = (3 - 0)^2 \times \frac{1}{6} + (-0.6 - 0)^2 \times \frac{5}{6} = 1.8.$$

You might wonder why $E(|X - \mu|)$ is not used as measure for the spread of X. One answer is that this measure has not such nice mathematical properties as the mean square of the deviations, but the use of the variance over the mean absolute error is really justified by the normal probability distribution and the central limit theorem, which will be discussed in Section 3.2.

The notation $\sigma^2(X)$ is often used for var(X). Alternatively, var(X) can be calculated as

$$\boxed{\text{var}(X) = E(X^2) - \mu^2,}$$

as can be seen by writing $(X - \mu)^2$ as $X^2 - 2\mu X + \mu^2$ and using the linearity property of expectation. Also, using this property and the definition of variance, you can readily verify that

$$\boxed{\text{var}(aX + b) = a^2\text{var}(X) \text{ for any constants } a \text{ and } b.}$$

By the substitution rule, the variance of a discrete random variable X with $a_1, a_2, \ldots, a_r$ as possible values can be calculated as

$$\text{var}(X)=\sum_{j=1}^{r}(a_j - \mu)^2 P(X = a_j) \text{ or } \text{var}(X)=\sum_{j=1}^{r} a_j^2 P(X = a_j) - \mu^2.$$

The variance of a random variable X does not have the same units (e.g., dollar or cm) as $E(X)$. For example, if X is expressed in dollars, then $\text{var}(X)$ has (dollars)^2 as dimension. Therefore, one often uses the *standard deviation* that is defined as the square root of the variance. The standard deviation of X is denoted by $\sigma(X)$, that is,

$$\boxed{\sigma(X) = \sqrt{\text{var}(X)}.}$$

As an illustration, suppose that X is the number of points to be obtained in a single roll of a die. Since $E(X) = 3.5$, the variance and standard deviation of X are $\sum_{j=1}^{6}(j - 3.5)^2 \frac{1}{6} = \frac{35}{12}$ and $\sqrt{\frac{35}{12}}$.

Example 2.16. Joe and his friend bet every week whether the Dow Jones index will have risen at the end of the week or not. Both put $10 in the pot. Joe observes that his friend is just guessing and is making his choice by the toss of a fair coin. Joe asks his friend if he could contribute $20 to the pot and submit his guess together with that of his brother. The friend agrees. In each week, however, Joe's brother submits a prediction opposite to that of Joe. If there is only one correct prediction, the entire pot goes to that prediction. If there is more than one correct prediction, the pot is split evenly between the correct predictions. How favorable is the game to the brothers?

Solution. Let the random variable X denote the payoff to Joe and his brother in any given week. Either Joe or his brother will have a correct prediction. They win the $30 pot if Joe's friend is wrong;

otherwise, they share the pot with Joe's friend. The possible values of X are 30 and 15 dollars. Each of these two values is equally likely, since Joe's friend makes his prediction by the toss of a coin. Thus,

$$E(X) = 30 \times \frac{1}{2} + 15 \times \frac{1}{2} = 22.5 \text{ dollars.}$$

Joe and his brother have an expected profit of $E(X - 20) = 2.5$ dollars, each week. To obtain $\sigma(X)$, you first calculate

$$E(X^2) = 900 \times \frac{1}{2} + 225 \times \frac{1}{2} = 562.5 \text{ (dollars)}^2.$$

This gives $\text{var}(X) = 562.5 - 22.5^2 = 56.25$ (dollars)^2. Thus,

$$\sigma(X) = \sqrt{56.25} = 7.5 \text{ dollars.}$$

Since $\sigma(X - 20) = \sigma(X)$, the standard deviation of the profit $X - 20$ for Joe and his brother is also equal to 7.5 dollars.

Problem 2.32. You roll a die. If a 6 six appears, you win \$10. If not, you roll the die again. If a 6 appears the second time, you win \$5. If not, you win nothing. What are the expected value and the standard deviation of your winnings? (answer: \$2.36 and \$3.82)

Problem 2.33. Consider again Problem 2.30. What are the expected value and standard deviation of X? (answer: 4.472 and 1.404)

Problem 2.34. Consider again Problem 2.31. What are the expected value and the standard deviation of the sum of the scores on the rolled dice? (answer: 3.5 and 3.001)

Problem 2.35. Investment A has a 0.8 probability of a \$2 000 profit and a 0.2 probability of a \$3 000 loss. Investment B has a 0.2 probability of a \$5 000 profit and a 0.8 probability of a zero profit. Verify that both investments have the same expected value and the same standard deviation for the net profit. (answer: expected value is \$1 000 and standard deviation is \$2 000)

Problem 2.36. There are four courses having 15, 20, 75, and 125 students. No student takes more than one course. Let the random variable X be the number of students in a randomly chosen class and Y be the number of students in the class of a randomly chosen student. Can you explain beforehand why $E(Y)$ is larger than $E(X)$? What are $E(X)$ and $E(Y)$? (answer: 58.750 and 93.085) What are the coefficients of variation (or relative standard deviations) $\sigma(X)/E(X)$ and $\sigma(Y)/E(Y)$? (answer: 0.764 and 0.415)

Problem 2.37. You throw darts at a circular target on which two concentric circles of radius 1 cm and 3 cm are drawn. The target itself has a radius of 5 cm. You receive 15 points for hitting the target inside the smaller circle, 8 points for hitting the middle annular region, and 5 points for hitting the outer annular region. The probability of hitting the target at all is 0.75. If the dart hits the target, then the hitting point is a random point on the target. Let the random variable X be the number of points scored on a single throw of the dart. What is the expected value of X? (answer: 4.77)

Problem 2.38. Shuffle an ordinary deck of 52 playing cards. Then turn up the cards from the top until the first ace appears. What is the expected number of cards to be turned over? (answer: 10.6)

Problem 2.39. A bowl has 10 white and 2 red balls. You pick m balls at random, where m can be chosen at your discretion. If each picked ball is white, you win \$$m$; otherwise, you win nothing. What value of m maximizes your expected winnings? (answer: $m = 4$)

Problem 2.40. You have five distinct pairs of socks in a drawer. The socks are not folded in pairs. You pick socks out of the drawer, one at a time and at random. What are the expected value and the standard deviation of the number of socks you must pick out of the drawer in order to get a matching pair? (answer: 4.06 and 1.19)

Problem 2.41. In a barn, 100 chicks sit peacefully in a circle. Suddenly, each chick randomly pecks the chick immediately to its left or right. What is the expected value of the number of unpicked chicks? (answer: 25)

Problem 2.42. What is the expected number of different values that come up when six fair dice are rolled? (answer: 3.991)

Problem 2.43. In the lotto 6/42 a player chooses six different numbers from $1, \ldots, 42$. Suppose players have filled in 5 million tickets with random picks. What is the expected value of the number of different six-number combinations filled in? (answer: $3\,223\,398$)

Problem 2.44. A random variable X is said have a *discrete uniform distribution* with integer-valued parameters a and b $(> a)$ if

$$P(X = k) = \frac{1}{b - a + 1} \qquad \text{for } k = a, a + 1, \ldots, b.$$

Verify that $E(X) = \frac{1}{2}(b - a)$ and $\sigma^2(X) = \frac{1}{12}[(b - a + 1)^2 - 1]$.

Problem 2.45. A random variable X is said to have a *Bernoulli distribution* with parameter p $(0 < p < 1)$ if

$$P(X = 1) = p \quad \text{and} \quad P(X = 0) = 1 - p.$$

Verify that $E(X) = p$ and $\sigma(X) = \sqrt{p(1 - p)}$.

Problem 2.46. A random variable X is said to have a *geometric distribution* with parameter p $(0 < p < 1)$ if

$$P(X = k) = (1 - p)^{k-1} p \qquad \text{for } k = 1, 2, \ldots$$

Verify that $E(X) = \frac{1}{p}$ and $\sigma(X) = \frac{1}{p}\sqrt{1 - p}$.

Problem 2.47. Verify the formulas $E(N) = \sum_{k=0}^{\infty} P(N > k)$ and $E[N(N - 1)] = \sum_{k=0}^{\infty} 2kP(N > k)$ for a non-negative, integer-valued random variable N.

Problem 2.48. What is the expected number of boxes that must be purchased in order to get a complete set of cards in the coupon collector's problem from Example 2.15? (answer: 224.96)

Problem 2.49. What is the expected number of rolls of a fair die it takes to see all six sides of the die at least once? (answer: 14.7)

Problem 2.50. You play a game in which you can pick a random number from 1 to 25 as often as you wish. Each pick costs you one dollar. If you decide to stop, you get paid in dollars your last picked number. What strategy maximizes your expected net payoff? (answer: stop if your picked number is ≥ 19)

Problem 2.51. On a game show, you can bet on one of the numbers 1 to 100. Then, a random number is generated from 1 to 100. If your guess is less than the randomly chosen number, you win in dollars the square of your guess; otherwise, you win nothing. What number should you guess to maximize your expected winning? (answer: 67)

2.7 Independent random variables and the square root law

As you have seen in Section 2.6, it is always true that $E(X + Y) = E(X) + E(Y)$ for any two random variables X and Y. A similar result for the variance is in general not true. You can see this from the example with $P(X = 1) = P(X = -1) = \frac{1}{2}$ and $Y = -X$. In this example, $\text{var}(X + Y) = 0$ and $\text{var}(X) = \text{var}(Y) = 1$ (verify!). The reason that $\text{var}(X + Y)$ is not equal to $\text{var}(X) + \text{var}(Y)$ is that X and Y are not independent of each other. Two random variables X and Y are said to be *independent* of each other if

$$\boxed{P(X \leq x \text{ and } Y \leq y) = P(X \leq x)P(Y \leq y) \text{ for all } x \text{ and } y.}$$

For discrete random variables X and Y, an alternative definition of independence is $P(X = x \text{ and } Y = y) = P(X = x)P(Y = y)$ for all x and y. Then the following result holds

$$\boxed{\text{var}(aX + bY) = a^2\text{var}(X) + b^2\text{var}(Y) \text{ for independent } X \text{ and } Y,}$$

where a and b are constants. This result is true for any type of random variables. For the special case of discrete random variables X and Y, a proof is given in Section 2.10. More generally, for independent $X_1, \ldots, X_n$ and constants $c_1, \ldots, c_n,$[14]

$$\boxed{\text{var}(c_1 X_1 + \cdots + c_n X_n) = c_1^2 \, \text{var}(X_1) + \cdots + c_n^2 \, \text{var}(X_n).}$$

---

[14]In general, random variables $X_1, \ldots, X_n$ are said to be independent if $P(X_1 \leq x_1 \text{ and } \ldots \text{ and } X_n \leq x_n) = P(X_1 \leq x_1) \cdots P(X_n \leq x_n)$ for all $x_1, x_2, \ldots, x_n$.

As an illustration, what is the standard deviation of the sum of one roll of two dice? Let X be the outcome of the first die and Y of the second die. The sum of one roll of two dice is $X + Y$. As calculated before, $\text{var}(X) = \text{var}(Y) = \frac{35}{12}$. The random variables X and Y are independent of each other. Then $\text{var}(X+Y) = \text{var}(X)+\text{var}(Y) = \frac{35}{6}$. Thus, the standard deviation of the sum of one roll of two dice is $\sigma(X + Y) = \sqrt{\frac{35}{6}} \approx 2.415$ points.

The square root law for the standard deviation

Let $X_1, \ldots, X_n$ be independent random variables each having standard deviation σ. Then, by $\text{var}\left(\sum_{k=1}^{n} X_k\right) = \sum_{k=1}^{n} \text{var}(X_k) = n\sigma^2$ and $\text{var}\left(\frac{1}{n}\sum_{k=1}^{n} X_k\right) = \frac{1}{n^2}\sum_{k=1}^{n} \text{var}(X_k) = \frac{\sigma^2}{n}$, you have

$$\sigma\left(\sum_{k=1}^{n} X_k\right) = \sigma\sqrt{n} \quad \text{and} \quad \sigma\left(\frac{1}{n}\sum_{k=1}^{n} X_k\right) = \frac{\sigma}{\sqrt{n}}.$$

This is called the *square root law* (or the $\sqrt{n}$-law) for the standard deviation. It is an extremely important result in probability and statistics. In Figure 2 an experimental demonstration of the $\sqrt{n}$-law is given. For each of the values $n = 1, 4, 16$, and 64, one hundred random outcomes of the so-called sample mean $\frac{1}{n}\sum_{k=1}^{n} X_k$ are simulated for the case that the X_k have a same probability distribution with expected value 0 and standard deviation 1 (the chosen distribution is the standard normal distribution, which will be discussed in Section 3.4). You see from the figure that the bandwidths within which the simulated outcomes lie are indeed reduced by a factor of about 2 when the sample sizes increase by a factor of 4.

The $\sqrt{n}$-law is sometimes called the De Moivre's equation, after Abraham de Moivre (1667–1754).[15] This formula had an immediate impact on methods used to inspect gold coins struck at the London

---

[15]The French-horn Abraham de Moivre was the leading probabilist of the eighteenth century and lived most of his life in England. The protestant De Moivre left France in 1688 to escape religious persecution. He was a good friend of Isaac Newton and supported himself by calculating odds for gamblers and insurers and by giving private lessons to students.

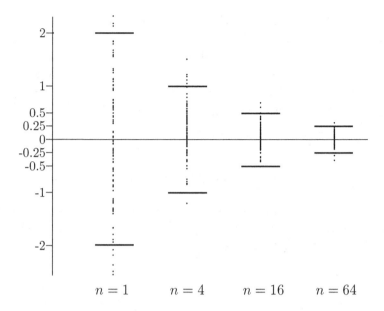

Figure 2: An illustration of the square root law.

Mint. The standard gold weight, per coin, was 128 grains (one grain was equal to 0.0648 gram), and the allowable deviation from this standard was $\frac{1}{400}$ of that amount, or 0.32 grains. A test case of 100 coins was periodically performed on coins struck, their total weight then being compared with the standard weight of 100 coins. The gold used in the striking of coins was the property of the king, who sent inspectors to discourage minting mischief. The royal watch dogs had traditionally allowed a deviation of $100 \times 0.32 = 32$ grains in the weight of 100 inspected coins. Directly after De Moivre's publication of the square root formula in 1733, the allowable deviation in the weight of 100 coins was changed to $\sqrt{100} \times 0.32 = 3.2$ grains; alas for the English monarchy, previous ignorance of the square root formula had cost them a fortune in gold.

The square root law has many applications, providing explanation, for example, for why city or hospital size is important for measuring crime statistics or death rates after surgery. Small hospitals, for example, are more likely than large ones to appear at the top or bottom of ranking lists. This makes sense if you consider that, when tossing a fair coin, the probability that more than 70%, or less than 30%, of the tosses will turn up heads is much larger for 10 coin tosses, than for 100. The smaller the numbers, the larger the chance fluctuations!

Example 2.17. 'Unders and Overs' is a popular game formerly played during open house events at American schools, for the purpose of adding money to the school coffers. The game is played with two dice and a playing board divided into three sections: 'Under 7', '7', and 'Over 7'. The two dice are rolled, and players place chips on one or more of the three sections. Chips may be placed on the game board for 1 dollar apiece. For every chip placed in the 'Under 7' section, the payoff is 2 dollars if the total number of points rolled with the dice is less than 7. The payoff is the same for every chip in the 'Over 7' section if the total number of points is higher than 7. The payoff is 5 dollars for each chip placed on '7' if the total number of points is 7. A popular strategy is to place 1 chip on each of the three sections. Suppose that 500 rounds of the game are played using this strategy. In each round there is a single player. What are the expected value and standard deviation of the net amount taken in by the school as a result of the 500 bets?

Solution. Let the random variable X be the net profit of the school in a single play of the game. The random variable X can take on the two values \$1 and $-$\$2. The net profit is $3 - 5 = -2$ if the sum of the dice is 7 and is $3 - 2 = 1$ otherwise. Using the results of Example 2.10, you find $P(X = -2) = \frac{6}{36}$ and $P(X = 1) = 1 - \frac{6}{36} = \frac{30}{36}$. Thus,

$$E(X) = -2 \times \frac{6}{36} + 1 \times \frac{30}{36} = \frac{1}{2}.$$

The alternative definition $\sigma^2(X) = E(X^2) - \big(E(X)\big)^2$ is used to calculate $\sigma(X)$. Since $E(X^2) = 4 \times \frac{6}{36} + 1 \times \frac{30}{36} = \frac{3}{2}$, you get $\sigma^2(X) = \frac{3}{2} - (\frac{1}{2})^2 = \frac{5}{4}$ and so

$$\sigma(X) = \frac{1}{2}\sqrt{5}.$$

The total net profit of the school is $X_1 + \cdots + X_{500}$ dollars, where X_i is the net profit of the school in the ith round. The random variables

$X_1, \ldots, X_{500}$ are each distributed as X and are independent of each other. Using the linearity property of expectation,

$$E(X_1 + \cdots + X_{500}) = 500 \times \frac{1}{2} = 250 \text{ dollars.}$$

By the $\sqrt{n}$-law for the standard deviation,

$$\sigma(X_1 + \cdots + X_{500}) = \frac{1}{2}\sqrt{5} \times \sqrt{500} = 25 \text{ dollars.}$$

Problem 2.52. A fair die will be thrown 100 times. What are the expected value and the standard deviation of the average number of points per throw? (answer: 3.5 and 0.171)

Problem 2.53. The Mang Kung dice game is played with six non-traditional dice. Each of the six dice has five blank faces and one face marked with one of the numbers 1 up to 6 such that no two dice have the same number. What are the expected value and the standard deviation of the total number of points showing up when the six dice are rolled? (answer: 3.5 and 3.555)

Problem 2.54. Consider Problem 2.48 again. Use results from Problem 2.46 to calculate the standard deviation of the number of purchases needed to get a complete set of cards. (answer: 61.951)

Problem 2.55. Consider Problem 2.49 again. What is the standard deviation of the number of rolls needed to get all six possible outcomes? (answer: 6.244)

2.8 Generating functions

Counting was used in Example 2.12 to calculate the probability mass function of the sum of the upturned faces when rolling two dice. The counting approach becomes very tedious when asking about three dice, not to mention five or ten dice. A kind of magic approach to handle these cases is the generating function approach. Generating functions were introduced by the Swiss genius Leonhard Euler (1707–1783) to facilitate calculations in counting problems. The idea is very simple and illustrates that simple ideas are often the best.

The generating function $G_X(z)$ of a non-negative, integer-valued random variable X is defined by the power series

$$G_X(z) = \sum_{k=0}^{\infty} P(X = k)z^k \quad \text{for } |z| \leq 1.$$

The generating function $G_X(z)$ uniquely determines the probability mass function of X: taking the rth derivative of $G_X(z)$ at $z = 0$ gives $r!P(X = r)$ (verify). The derivative of $G_X(z)$ at $z = 1$ is equal to $\sum_{k=0}^{\infty} k\,P(X = k)$ and so the expected value of X can be calculated as

$$E(X) = G_X'(1).$$

Similarly, taking the second derivative of $G_X(z)$ at $z = 1$ leads to

$$E[X(X - 1)] = G_X''(1).$$

A useful representation can be given for $G_X(z)$. By the substitution rule, $E(z^X) = \sum_{k=0}^{\infty} z^k\,P(X = k)$ and so

$$G_X(z) = E(z^X) \quad \text{for } |z| \leq 1.$$

A very important result is that

$$G_{X+Y}(z) = G_X(z)G_Y(z) \quad \text{for } |z| \leq 1.$$

for independent random variables X and Y on the non-negative integers. That is, the generating function of the sum of independent random variables is the product of the generating functions of the individual random variables. This result can be explained as follows. First, it is noted that $f(X)$ and $g(Y)$ are independent random variables for any functions $f(x)$ and $g(y)$. Taking this fact for granted, it follows that the random variables z^X and z^Y are independent for any $|z| \leq 1$. In Section 2.10 it will be proved that $E(VW) = E(V)E(W)$ for independent random variables V and W. Thus, for any $|z| \leq 1$,

$$G_{X+Y}(z) = E(z^{X+Y}) = E(z^X z^Y) = E(z^X)E(z^Y),$$

and so $G_{X+Y}(z) = G_X(z)G_Y(z)$, as was to be verified.

Example 2.18. A random variable X with the probability mass function

$$P(X = k) = e^{-\lambda}\frac{\lambda^k}{k!} \quad \text{for } k = 0, 1, \ldots$$

is said to have a *Poisson distribution* with parameter λ, where λ is a positive number. What is the generating function of X? What is $E(X)$? Suppose that X_1 and X_2 are independent random variables having Poisson distributions with respective parameters λ_1 and λ_2. What is the probability distribution of the sum $X_1 + X_2$?

Solution. The generating function of a random variable X having a Poisson distribution with parameter λ is

$$G_X(z) = \sum_{k=0}^{\infty} z^k P(X = k) = \sum_{k=0}^{\infty} e^{-\lambda}\frac{\lambda^k}{k!}z^k = e^{-\lambda}\sum_{k=0}^{\infty}\frac{(\lambda z)^k}{k!} = e^{-\lambda}e^{\lambda z}$$

$$= e^{-\lambda(1-z)} \quad \text{for } |z| \leq 1.$$

The generating function $G_X(z) = e^{-\lambda(1-z)}$ uniquely determines the Poisson distribution with parameter λ. By $G_X'(1) = \lambda$, the expected value of X is λ. To answer the third question, note that

$$G_{X_1+X_2}(z) = E\big(z^{X_1+X_2}\big) = E\big(z^{X_1}\big)E\big(z^{X_2}\big),$$

where the latter equality uses the independence of X_1 and X_2. Thus,

$$G_{X_1+X_2}(z) = e^{-\lambda_1(1-z)}e^{-\lambda_2(1-z)} = e^{-(\lambda_1+\lambda_2)(1-z)} \quad \text{for } |z| \leq 1.$$

The function $e^{-(\lambda_1+\lambda_2)(1-z)}$ is the generating function of a Poisson distributed random variable with parameter $\lambda_1 + \lambda_2$. This shows that the sum of two independent random variables having Poisson distributions with parameters λ_1 and λ_2 is Poisson distributed with parameter $\lambda_1 + \lambda_2$.

Generating functions are very useful for computational purposes. As an example, let us apply the generating function method to the calculation of the probability mass function of the sum of the upturned faces when rolling three dice. For $i = 1, 2, 3$, let X_i be the upturned face value of the ith die. The generating function of X_i is

$$G_{X_i}(z) = \frac{1}{6}z + \frac{1}{6}z^2 + \frac{1}{6}z^3 + \frac{1}{6}z^4 + \frac{1}{6}z^5 + \frac{1}{6}z^6 \quad \text{for } i = 1, 2, 3.$$

Since X_1, X_2, and X_3 are independent, the generating function of the sum of the upturned face values is given by

$$G_{X_1+X_2+X_3}(z) = \left(\frac{1}{6}z + \frac{1}{6}z^2 + \frac{1}{6}z^3 + \frac{1}{6}z^4 + \frac{1}{6}z^5 + \frac{1}{6}z^6\right)^3 \quad \text{for } |z| \le 1.$$

Using standard software for calculating the product of polynomials,

$$\begin{aligned}
G_{X_1+X_2+X_3}(z) &= \frac{1}{216}z^3 + \frac{1}{72}z^4 + \frac{1}{36}z^5 + \frac{5}{108}z^6 + \frac{5}{72}z^7 + \frac{7}{72}z^8 \\
&+ \frac{25}{216}z^9 + \frac{1}{8}z^{10} + \frac{1}{8}z^{11} + \frac{25}{216}z^{12} + \frac{7}{72}z^{13} \\
&+ \frac{5}{72}z^{14} + \frac{5}{108}z^{15} + \frac{1}{36}z^{16} + \frac{1}{72}z^{17} + \frac{1}{216}z^{18}.
\end{aligned}$$

The probability mass function of the sum can be directly read off from this expansion. The coefficient of z^k gives the probability that the sum $X_1 + X_2 + X_3$ takes on the value k for $k = 3, \ldots, 18$. Can you explain why the probability for the sum $21 - k$ is the same as the probability for the sum k for $k = 3, \ldots, 10$?

Problem 2.56. Use the generating function method to find the standard deviation of a Poisson distributed random variable with parameter λ. (answer: $\sqrt{\lambda}$)

Problem 2.57. You have 8 symmetric six-sided dice. Five of these dice have the number six on two of the faces and the other three have the number six on three of the faces. What is the probability mass function of the number of sixes appearing when rolling the 8 dice? (answer: $\left(\frac{4}{243}, \frac{22}{243}, \frac{52}{243}, \frac{23}{81}, \frac{25}{108}, \frac{77}{648}, \frac{73}{1944}, \frac{13}{1944}, \frac{1}{1944}\right)$)

Problem 2.58. Let $X_1, \ldots, X_5$ be independent Bernoulli variables with $P(X_i = 1) = \frac{1}{i+1}$ and $P(X_i = 0) = 1 - \frac{1}{i+1}$ for $i = 1, \ldots, 5$. What is the probability mass function of the sum $X_1 + \cdots + X_5$? (answer: $\left(\frac{1}{6}, \frac{137}{360}, \frac{5}{16}, \frac{17}{144}, \frac{1}{48}, \frac{1}{720}\right)$)

2.9 Inequalities and the law of large numbers

This section discusses famous inequalities in probability and two versions of the law of arge numbers. The inequalities include Markov's inequality and Chebyshev's inequality. The latter inequality will be used to prove the weak law of large numbers, which goes back to Jakob Bernoulli (1654–1705). The proof of the strong law of numbers will not be given, but this law will be illustrated with the derivation of the Kelly betting fraction in a repeatable gambling game in which the player has an advantage.

Markov's inequality

For a *non-negative* random variable Y with finite expected value, Markov's inequality states that

$$P(Y \geq a) \leq \frac{E(Y)}{a} \qquad \text{for any constant } a > 0.$$

The proof is simple. For fixed $a > 0$, let the indicator variable I be equal to 1 if $Y \geq a$ and 0 otherwise. Then, $E(I) = P(Y \geq a)$. By the non-negativity of Y, you have that $Y \geq aI$, and so

$$E(Y) \geq aE(I) = aP(Y \geq a),$$

which gives the inequality. Beauty in simplicity! The inequality was proved by the Russian mathematician Andrey A. Markov (1856–1922) in 1889 and it is the basis for many other useful inequalities in probability, including the so-called Chernoff bounds which we briefly touch on in Problem 2.60 below.

Chebyshev's inequality

For any random variable X having finite expected value μ and finite variance σ^2, Chebyshev's inequality states that

$$P(|X - \mu| \geq c) \leq \frac{\sigma^2}{c^2} \qquad \text{for any constant } c > 0.$$

This inequality is named after the Russian mathematician Pafnuty L. Chebyshev (1821–1894) who proved the inequality in 1867. A simpler proof was later given by his famous student Andrey Markov: taking $Y = (X - \mu)^2$ and $a = c^2$ in Markov's inequality, you get

$$P(|X - \mu| \geq c) = P((X - \mu)^2 \geq c^2) \leq \frac{E(X - \mu)^2}{c^2} = \frac{\sigma^2}{c^2}.$$

Law of large numbers

There are two main versions of the law of large numbers. They are called the *weak* and *strong* laws of large numbers. Both laws deal with the behavior of the sample mean $\frac{1}{n}\sum_{k=1}^{n} X_k$ for large n, where $X_1, X_2, \ldots, X_n$ are independent random variables having a same probability distribution with finite mean μ. The assumption of a finite variance σ^2 is not required for the laws, but is often used because it makes the proofs easier and shorter. The weak law of large numbers essentially states that for any nonzero specified margin, no matter how small, there is a high probability that the sample mean for large n will be close to its expected value μ within the margin. That is,

$$\lim_{n\to\infty} P\left(\left|\frac{1}{n}\sum_{k=1}^{n} X_k - \mu\right| \geq \epsilon\right) = 0 \quad \text{for any } \epsilon > 0.$$

A simple proof can be given when it is assumed that the variance σ^2 of the random variables X_k is finite. Noting that the sample mean $\frac{1}{n}\sum_{k=1}^{n} X_k$ has expected value μ and standard deviation $\sigma/\sqrt{n}$, it follows from Chebyshev's inequality that

$$P\left(\left|\frac{1}{n}\sum_{k=1}^{n} X_k - \mu\right| \geq \epsilon\right) \leq \frac{(\sigma/\sqrt{n})^2}{\epsilon^2} \quad \text{for any } n \geq 1 \text{ and } \epsilon > 0.$$

By letting $n \to \infty$, the result follows.

The strong law of large numbers states that with probability 1 the sequence of sample means $\frac{1}{n}\sum_{k=1}^{n} X_k$ converges to μ as n becomes very large. That is,

$$P\left(\lim_{n\to\infty} \frac{1}{n}\sum_{k=1}^{n} X_k = \mu\right) = 1.$$

In other words, in practice you always get a realization ω of the sequence $X_1, \ldots, X_n$ such that the average value $\frac{1}{n} \sum_{k=1}^{n} X_k(\omega)$ converges to μ as n becomes very large. In accordance with our intuition, the strong law of large numbers 'guarantees' stable long-term results for random events. For example, a casino may lose money in a few spins of the roulette wheel, but its earnings will tend to a predictable percentage over a large number of spins. Not only the profits of casinos are based on the strong law, but the profits of insurance companies as well. Also, estimating probabilities by Monte Carlo simulation relies on the strong law of large numbers.

2.9.1 Kelly formula in gambling and investment

In this subsection, the law of large numbers will be used to derive the famous Kelly betting formula in gambling and investment.

Example 2.19. Consider the situation in which you can repeatedly make bets in a particular game with a single betting object. The outcomes of the successive bets are independent of each other. For every dollar bet, you receive w_1 dollars back with probability p and w_2 dollars with probability $1 - p$, where $0 < p < 1$, $w_1 > 1$, and $0 \leq w_2 < 1$. The key assumption is that the game is favorable to the player, that is, $pw_1 + (1 - p)w_2 > 1$. Also, it is assumed that $pw_1 + (1 - p)w_2 - 1 < (w_1 - 1)(1 - w_2)$. How should you bet to manage your bankroll in a good way in the long run?

Solution. If you bet your entire bankroll each time to maximize the expected value of your winnings, ruin beckons. If you bet too little, the advantage is squandered. The optimal bet size is found by maximizing the expected value of the logarithm of wealth, which is equivalent to maximizing the expected geometric growth rate of your bankroll. It is best to bet the same fixed proportion α^* of your bankroll each time according to the *Kelly betting formula*

$$\alpha^* = \frac{pw_1 + (1 - p)w_2 - 1}{(w_1 - 1)(1 - w_2)}.$$

Note that, by the assumptions made, $0 < \alpha^* < 1$. In the special case of $w_2 = 0$, the Kelly formula reduces to

$$\alpha^* = \frac{pw_1 - 1}{w_1 - 1},$$

which can be interpreted as the ratio of the expected net gain per staked dollar and the payoff odds.

The derivation of the Kelly formula goes as follows. The strategy is to bet a fixed fraction α of your current bankroll each time, where $0 < \alpha < 1$. Letting V_0 be your starting bankroll, define the random variable V_n as the size of your bankroll after n bets. Let the random variable $N_{n,k}$ denote how many of the first n bets result in a payoff w_k for $k = 1, 2$. Then V_n can be represented as

$$V_n = (1 - \alpha + \alpha w_1)^{N_{n,1}} \times (1 - \alpha + \alpha w_2)^{N_{n,2}} V_0 \quad \text{for } n = 1, 2, \ldots .$$

The growth factor G_n of your bankroll V_n after n bets is defined through $V_n = e^{nG_n} V_0$, or equivalently, $G_n = \frac{1}{n} \ln(V_n/V_0)$. This gives

$$G_n = \frac{N_{n,1}}{n} \ln\left(1 - \alpha + \alpha w_1\right) + \frac{N_{n,2}}{n} \ln\left(1 - \alpha + \alpha w_2\right) \quad \text{for } n = 1, 2, \ldots .$$

The random variable $N_{n,1}$ can be written as the sum of n independent 0–1 indicator variables, where each variable is equal to 1 with probability p and equal to 0 with probability $1 - p$, and thus has expected value p. Thus, by the law of large numbers, $N_{n,1}/n$ converges to p with probability 1 as $n \to \infty$. By the same argument, $N_{n,2}/n$ converges to $1 - p$ with probability 1 as $n \to \infty$. Thus, the long-run growth rate of your bankroll is equal to

$$\lim_{n \to \infty} G_n = p \ln(1 - \alpha + \alpha w_1) + (1 - p) \ln(1 - \alpha + \alpha w_2) \text{ with probability 1.}$$

Putting the derivative of $g(\alpha) = p \ln(1 - \alpha + \alpha w_1) + (1 - p) \ln(1 - \alpha + \alpha w_2)$ equal to 0 and using the fact that the function $g(\alpha)$ is concave in α, you get after a little algebra that $g(\alpha)$ takes on its absolute maximum for $\alpha = \alpha^*$, verifying the Kelly formula.

Problem 2.60. Use Markov's inequality to verify for a generally distributed random variable X the generic *Chernoff bound*

$$P(X \geq c) \leq \frac{E\left(e^{tX}\right)}{e^{tc}} \quad \text{for } t \geq 0 \text{ and any } c.$$

2.10 Additional material

This section first proves the properties that were given in the Sections 2.6 and 2.7 for the expected value and variance. It then discusses the concepts of covariance, correlation, conditional expectation, and the classification tool of logistic regression in data science.

The proofs for the properties of the expected value and variance will be given for the case that X and Y are discrete random variables that can take on only a finite number of values. Let I be the set of possible values of X and J be the set of possible values of Y. For the moment, you are asked to take for granted the following result that will be proved below: for any function $g(x, y)$, the expected value of the random variable $g(X, Y)$ is given by the two-dimensional substitution rule

$$E\big[g(X,Y)\big] = \sum_{x \in I} \sum_{y \in J} g(x,y) P(X = x \text{ and } Y = y).$$

A double sum $\sum_{i=1}^{n} \sum_{j=1}^{m} a_{ij}$ should be read as $\sum_{i=1}^{n} (a_{i1} + \cdots + a_{im})$. You can always interchange the order of summation when there are finitely many terms: $\sum_{i=1}^{n} \sum_{j=1}^{m} a_{ij} = \sum_{j=1}^{m} \sum_{i=1}^{n} a_{ij}$. The following properties can now be easily proved:

Property 1. $E(aX + bY) = aE(X) + bE(Y)$ for constants a and b.

Property 2. $E(XY) = E(X)E(Y)$ if X and Y are independent of each other.

Property 3. $\text{var}(aX + bY) = a^2 \text{var}(X) + b^2 \text{var}(Y)$ for any constants a and b if X and Y are independent of each other.

Proof of Property 1. Using the basic result for $E\big(g(X,Y)\big)$ for $g(x,y) = ax + by$, you get that $E(aX + bY)$ is equal to

$$\sum_{x \in I} \sum_{y \in J} (ax + by) P(X = x \text{ and } Y = y)$$

$$= \sum_{x \in I} \sum_{y \in J} ax \, P(X = x \text{ and } Y = y) + \sum_{x \in I} \sum_{y \in J} by \, P(X = x \text{ and } Y = y)$$

$$= a \sum_{x \in I} x \sum_{y \in J} P(X = x \text{ and } Y = y) + b \sum_{y \in J} y \sum_{x \in I} P(X = x \text{ and } Y = y),$$

where the order of summation is interchanged in the second term of the last equation. Next you use the formula

$$P(X = x) = \sum_{y \in J} P(X = x \text{ and } Y = y).$$

This formula is a direct consequence of Axiom 3 in Section 2.1. A similar formula applies to $P(Y = y)$. This gives

$$E(aX+bY) = a \sum_{x \in I} xP(X = x) + b \sum_{y \in J} yP(Y = y) = aE(X)+bE(Y).$$

Proof of Property 2. The definition of independent random variables X and Y is $P(X = x \text{ and } Y = y) = P(X = x)P(Y = y)$ for all possible values x and y. Next, by applying the basic result for $E(g(X, Y))$ with $g(x, y) = xy$, you find that $E(XY)$ is equal to

$$\sum_{x \in I} \sum_{y \in J} xyP(X = x \text{ and } Y = y) = \sum_{x \in I} \sum_{y \in J} xyP(X = x)P(Y = y)$$
$$= \sum_{x \in I} xP(X = x) \sum_{y \in J} yP(Y = y) = E(X)E(Y).$$

Proof of Property 3. For ease of notation, write $E(X)$ as μ_X and $E(Y)$ as μ_Y. Using the alternative definition $\text{var}(V) = E(V^2) - \mu^2$ for the variance of a random variable V with expected value μ and using Property 1, you get

$$\text{var}(aX + bY) = E[(aX + bY)^2] - (E(aX + bY))^2$$
$$= a^2 E(X^2) + 2abE(XY) + b^2 E(Y^2) - (a\mu_X + b\mu_Y)^2.$$

Next, you use the independence of X and Y. This gives $E(XY) = E(X)E(Y)$, by Property 2, and so $\text{var}(aX + bY)$ is equal to

$$a^2 E(X^2) + 2ab\mu_X \mu_Y + b^2 E(Y^2) - a^2 \mu_X^2 - 2ab\mu_X \mu_Y - b^2 \mu_Y^2$$
$$= a^2 [E(X^2) - \mu_X^2] + b^2 [E(Y^2) - \mu_Y^2] = a^2 \text{var}(X) + b^2 \text{var}(Y).$$

A special case of Property 3 is that $\text{var}(aX + b) = a^2 \text{var}(X)$ for any constants a and b. This follows by taking for Y a degenerate random variable with $P(Y = 1) = 1$ for which $\text{var}(Y) = 0$ (verify).

It remains to prove the formula $\sum_{(x,y)} g(x,y)P(X = x$ and $Y = y)$ for $E(g(X,Y))$. The trick is to define the random variable $Z = g(X,Y)$. Then, $\sum_{(x,y)} g(x,y)P(X = x$ and $Y = y)$ can be written as

$$\sum_z \left[\sum_{(x,y):g(x,y)=z} g(x,y)P(X = x \text{ and } Y = y) \right]$$

$$= \sum_z z \sum_{(x,y):g(x,y)=z} P(X = x \text{ and } Y = y) = \sum_z zP(Z = z)$$

$$= E(Z) = E[g(X,Y)].$$

Taking $Y = X$ and $g(x,y) = g(x)$, this result proves the substitution rule in Section 2.6 as special case.

2.10.1 Covariance and correlation

How to calculate $\text{var}(X + Y)$ if X and Y are not independent random variables? To do this, you need the concept of covariance. The *covariance* $\text{cov}(X,Y)$ of two random variables X and Y is defined by

$$\boxed{\text{cov}(X,Y) = E[(X - E(X))(Y - E(Y))].}$$

Note that $\text{cov}(X,X) = \text{var}(X)$. The formula for $\text{cov}(X,Y)$ can be written in the equivalent form

$$\boxed{\text{cov}(X,Y) = E(XY) - E(X)E(Y),}$$

by writing $(X - E(X))(Y - E(Y))$ as $XY - XE(Y) - YE(X) + E(X)E(Y)$ and using the linearity property of the expectation operator. It is straightforward to verify from the definition of covariance that

$$\boxed{\text{cov}(aX + c, bY + d) = ab\,\text{cov}(X,Y)}$$

for any constants a, b, c, and d. Also, using the fact that $E(XY) = E(X)E(Y)$ for independent X and Y (see Property 2), it follows that

$$\boxed{\text{cov}(X,Y) = 0 \quad \text{if } X \text{ and } Y \text{ are independent random variables.}}$$

However, the converse of this result is not always true. As a counterexample, let X take on the equally likely values -1 and 1, and let

$Y = X^2$. Then, by $E(X) = 0$ and $E(X^3) = 0$, you have $\text{cov}(X, Y) = E(X^3) - E(X)E(X^2) = 0$, but X and Y are dependent.

The following formula for the variance of any two random variables X and Y can now be formulated:

$$\text{var}(X + Y) = \text{var}(X) + 2\text{cov}(X, Y) + \text{var}(Y).$$

The proof of this result is a matter of some algebra. Put for abbreviation $\mu_X = E(X)$ and $\mu_Y = E(Y)$. Using the definition of variance and the linearity of the expectation operator, you have that

$$\text{var}(X + Y) = E[(X + Y - (\mu_X + \mu_Y))^2]$$
$$= E[(X - \mu_X)^2] + 2E[(X - \mu_X)(Y - \mu_Y)] + E[(Y - \mu_Y)^2]$$
$$= \text{var}(X) + 2\text{cov}(X, Y) + \text{var}(Y),$$

as was to be verified. The formula for $\text{var}(X + Y)$ can be extended to the case of finitely many random variables:

$$\text{var}(X_1 + \cdots + X_n) = \sum_{i=1}^{n} \text{var}(X_i) + 2 \sum_{i=1}^{n-1} \sum_{j=i+1}^{n} \text{cov}(X_i, X_j).$$

Another very important concept in statistics is the concept of correlation coefficient. The units of $\text{cov}(X, Y)$ are not the same as the units of $E(X)$ and $E(Y)$. Therefore, it is often more convenient to use the *correlation coefficient* of X and Y. This statistic is defined as

$$\rho(X, Y) = \frac{\text{cov}(X, Y)}{\sigma(X)\sigma(Y)},$$

provided that $\sigma(X) > 0$ and $\sigma(Y) > 0$. The correlation coefficient is a dimensionless quantity with the property that

$$-1 \leq \rho(X, Y) \leq 1.$$

The algebraic proof of this property is omitted. The random variables X and Y are said to be *uncorrelated* if $\rho(X, Y) = 0$. Independent random variables X and Y are always uncorrelated, since their covariance is zero.

However, the converse of this result is not always true. Nevertheless, $\rho(X, Y)$ is often used as a measure of the dependence of X and Y.

2.10.2 Conditional expectation

Let X and Y be two dependent random variables. The conditional expectation of X given that $Y = y$ is defined in an obvious way: take the expectation with respect to the probability distribution of X given that $Y = y$. The concept of conditional expectation will be discussed for the simplest case of discrete random variables, but all results also apply to the case of general random variables.

If X and Y are discrete random variables, the *conditional mass function* of X given that $Y = y$ is defined by

$$P(X = x \mid Y = y) = \frac{P(X = x \text{ and } Y = y)}{P(Y = y)},$$

in accordance with $P(A \mid B) = P(A \text{ and } B)/P(B)$. The *conditional expectation* of X given that $Y = y$ is defined by

$$E(X \mid Y = y) = \sum_x x\, P(X = x \mid Y = y).$$

The conditional expectation $E(X \mid Y = y)$ can be thought of as the best estimate of the random variable X given that $Y = y$. That is, the function $g(y)$ minimizing the squared error distance

$$E[(X - g(Y))^2]$$

can be shown to be $g(y) = E(X \mid Y = y)$. In words, if you observe y, the best prediction of the value of X is $E(X \mid Y = y)$. A simple heuristic argument can be given for this fact. Take one random variable Z and ask yourselves the question "what value of c minimizes $E[(Z - c)^2]$?" Since $E[(Z - c)^2] = E(Z^2) - 2cE(Z) + c^2$, taking the derivative with respect to c gives $-2E(Z) + 2c = 0$ and so $c = E(Z)$. In general, $E(X \mid Y = y)$ is a nonlinear function of y, but in some special cases it is a linear function, see also Section 3.9.

A very useful result is the *law of conditional expectation*:

$$E(X) = \sum_y E(X \mid Y = y)P(Y = y).$$

The proof is simple:

$$E(X) = \sum_x xP(X = x) = \sum_x x \sum_y P(X = x \text{ and } Y = y)$$

$$= \sum_x x \sum_y P(Y = y)P(X = x \mid Y = y)$$

$$= \sum_y P(Y = y) \sum_x xP(X = x \mid Y = y).$$

This shows that $E(X) = \sum_y E(X \mid Y = y)P(Y = y)$, by the definition of $E(X \mid Y = y)$. The computation of the unconditional expectation of X is often simplified by conditioning on an appropriately chosen random variable Y. In the context of the problem it is usually obvious how to choose Y. The law of conditional expectation can be seen as a generalization of the *law of conditional probability*:

$$\boxed{P(X = x) = \sum_y P(X = x \mid Y = y)\, P(Y = y) \quad \text{for all } x.}$$

Example 2.20. A bin contains 10 strings. You randomly choose two loose ends and tie them up. You continue until there are no more two free ends. What is the expected number of loops you get?

Solution. It is helpful to parameterize the problem by assuming n strings. Let the random variable X_n be the number of loops you get for the case of n strings. To find $E(X_n)$, let the conditioning variable Y_n be 1 if the first two ends you choose belong to the same string and let Y_n be zero otherwise. Then, by the law of conditional expectation,

$$E(X_n) = E(1 + X_{n-1})P(Y_n = 1) + E(X_{n-1})P(Y_n = 0),$$

where $P(Y_n = 1) = \frac{1}{2n-1}$ and $P(Y_n = 0) = 1 - \frac{1}{2n-1}$. This recursion can be rewritten as $E(X_n) = \frac{1}{2n-1} + E(X_{n-1})$, where $E(X_0) = 0$. This gives $E(X_n) = \sum_{k=1}^{n} \frac{1}{2k-1}$. In particular, $E(X_{10}) = 2.133$.

Recursive thinking and the law of conditional expectation are very useful in solving probability problems and often go together.

Problem 2.61. The returns of funds A and B have the same expected value, standard deviations σ_A and σ_B, and a negative correlation coefficient ρ_{AB}. A fraction f of your money is invested in A and a fraction $1 - f$ in B. Verify that the standard deviation of the portfolio's return is minimized by $f = \left(\sigma_B^2 - \sigma_A\sigma_B\rho_{AB}\right) / \left(\sigma_A^2 + \sigma_B^2 - 2\sigma_A\sigma_B\rho_{AB}\right)$.

Problem 2.62. The *least squares regression line* of a dependent variable Y with respect to X is defined by $y = \alpha + \beta x$, where the coefficients α and β minimize $E\left[(Y - (\alpha + \beta X))^2\right]$. Verify that the least squares regression line is $y = E(Y) + \rho(X, Y)\frac{\sigma(Y)}{\sigma(X)}(x - E(X))$.

Problem 2.63. In one roll of two fair dice, let X be the largest number rolled and Y be the sum of the roll. What is the value of the conditional expectation $E(X \mid Y = 10)$? (answer: $\frac{17}{3}$)

Problem 2.64. You randomly choose three different numbers from the numbers 1 to 100. Let X be the smallest of these numbers and Y be the largest. What is $E(X \mid Y = y)$? (answer: $\frac{1}{3}y$)

Problem 2.65. Suppose n cars start in a random order along an infinitely long one-lane highway. They are all going at different but constant speeds and cannot pass each other. If a faster car ends up behind a slower car, it must slow down to the speed of the slower car. Eventually the cars will clump up in traffic jams. What is the expected number of clumps of cars? (answer: $\sum_{k=1}^{n}\frac{1}{k}$) *Hint*: set up a recursion by conditioning on the position of the slowest car.

2.10.3 Logistic regression in data analysis

Logistic regression analyzes the relationship between multiple explanatory random variables $X_1, \ldots, X_s$ between which no strong correlation exists and a categorical dependent variable Y. This section only deals with the binary logistic regression model in which the response variable Y is a $0-1$ variable. As an illustrative example: how to predict the probability of approval of your application for a home mortgage given your credit score? Denoting by X your credit score and letting $Y = 1$ if your application is approved and $Y = 0$ otherwise, then logistic regression models the probability $p = P(Y = 1 \mid X = x)$ as

$$\ln\left(\frac{p}{1 - p}\right) = \theta_0 + \theta_1 x.$$

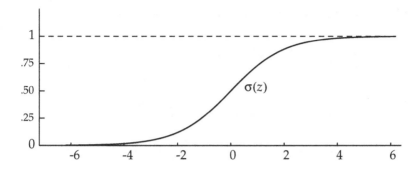

Figure 3: The sigmoid function.

This can be rewritten as $p/(1-p) = e^{\theta_0+\theta_1 x}$. Solving for p gives

$$P(Y = 1 \mid X = x) = \frac{1}{1 + e^{-(\theta_0+\theta_1 x)}}.$$

The so-called sigmoid function $\sigma(z)$ shows up in this formula, where

$$\sigma(z) = \frac{1}{1 + e^{-z}} \quad \text{for } -\infty < z < \infty.$$

This function is nearly linear around 0 and flattens towards the ends, see Figure 3. The function $\sigma(z)$ has the beautiful property $\sigma'(z) = \sigma(z)[1 - \sigma(z)]$. Since the variable Y is binary, $P(Y = 0 \mid X = x) = 1 - P(Y = 1 \mid X = x)$, and this leads to the useful representation

$$P(Y = y \mid X = x) = [\sigma(\theta_0+\theta_1 x)]^y \times [1-\sigma(\theta_0+\theta_1 x)]^{1-y} \quad \text{for } y = 0, 1.$$

The simple model with one explanatory variable can be extended to the case of multiple explanatory variables $X_1, \ldots, X_s$:

$$P(Y = 1 \mid X_j = x_j \text{ for } j = 1, \ldots, s) = \frac{1}{1 + e^{-(\theta_0+\theta_1 x_1 + \cdots + \theta_s x_s)}}.$$

How to estimate the parameters $\theta_0, \theta_1, \ldots, \theta_s$ from a sufficiently large set of data? Suppose that independent data points $(\mathbf{x}^{(i)}, y^{(i)})$ for $i = 1, \ldots, n$ are given, where $\mathbf{x}^{(i)} = (x_1^{(i)}, \ldots, x_s^{(i)})$. Using the shorthand notation $\boldsymbol{\theta} = (\theta_1, \ldots, \theta_s)$ and

$$\boldsymbol{\theta} \cdot \mathbf{x}^{(i)} = \theta_0 + \theta_1 x_1^{(i)} + \cdots + \theta_s x_s^{(i)},$$

the likelihood function of all data points is given by

$$L(\boldsymbol{\theta}) = \prod_{i=1}^{n} P\big(Y = y^{(i)} \mid (X_1, \ldots, X_s) = (x_1^{(i)}, \ldots, x_s^{(i)})\big)$$

$$= \prod_{i=1}^{n} [\sigma(\boldsymbol{\theta} \cdot \mathbf{x}^{(i)})]^{y^{(i)}} \times [1 - \sigma(\boldsymbol{\theta} \cdot \mathbf{x}^{(i)})]^{1-y^{(i)}}.$$

Taking the logarithm of both sides, you get the log likelihood function

$$LL(\boldsymbol{\theta}) = \sum_{i=1}^{n} y^{(i)} \ln[\sigma(\boldsymbol{\theta} \cdot \mathbf{x}^{(i)})] + \big(1 - y^{(i)}\big) \ln[1 - \sigma(\boldsymbol{\theta} \cdot \mathbf{x}^{(i)})].$$

To find the parameters θ_j that maximize the likelihood function, the classic gradient descent method is used. Maximizing the likelihood function is equivalent to minimizing the loss function $J(\boldsymbol{\theta}) = -LL(\boldsymbol{\theta})$. The loss function can be shown to be convex and thus has just one minimum. Therefore, gradient descent starting from any point is guaranteed to find the minimum. The algorithm requires the partial derivatives of $J(\boldsymbol{\theta})$. Using the property $\sigma'(z) = \sigma(z)[1 - \sigma(z)]$, it is a matter of simple algebra to get

$$\frac{\partial}{\partial \theta_j} J(\boldsymbol{\theta}) = -\sum_{i=1}^{n} [y^{(i)} - \sigma(\boldsymbol{\theta} \cdot \mathbf{x}^{(i)})] x_j^{(i)} \quad \text{for } j = 0, 1, \ldots, s,$$

where $x_0^{(i)} = 1$ for all i for a unified notation. Since gradient descent algorithm moves in the direction of the negative of the gradient, an iteration of the algorithm goes as follows:

$$\theta_j^{new} = \theta_j^{old} + \eta \sum_{i=1}^{n} [y^{(i)} - \sigma(\boldsymbol{\theta}^{old} \cdot \mathbf{x}^{(i)})] x_j^{(i)} \quad \text{for } j = 0, 1, \ldots, s.$$

The learning rate $\eta > 0$ is a flexible parameter that strongly influences the convergence of the algorithm. In large-scale data sets, convergence can be significantly improved by using stochastic gradient descent. This algorithm calculates the gradient for one or more observations picked at random instead of calculating the gradient for the entire data set.

Chapter 3

Useful Probability Distributions

In this chapter, the more common and important probability distributions are discussed in detail. The discrete binomial, hypergeometric and Poisson distributions are derived and illustrated with applications in Sections 3.1–3.3. The continuous normal, uniform, beta and exponential distributions and the Poisson process are covered in Sections 3.4–3.7. Other topics covered are the bivariate normal density, linear regression, and the chi-square test.

3.1 The binomial distribution

A random variable X with $0, 1, \ldots, n$ as possible values is said to have a *binomial distribution* with parameters n and p if

$$P(X = k) = \binom{n}{k} p^k (1 - p)^{n-k} \quad \text{for } k = 0, 1, \ldots, n,$$

where $\binom{n}{k}$ is the binomial coefficient. This distribution arises in probability problems that can be formulated within the framework of a sequence of physically independent trials, where each trial has the two possible outcomes 'success' (S) and 'failure' (F). The outcome success occurs with probability p and the outcome failure with probability $1 - p$. The random variable X defined as the total number of successes in n trials has a binomial distribution. This result is easily derived. The probability that a pre-specified sequence of k successes and $n - k$ failures will occur is $p^k (1 - p)^{n-k}$; for example, for $n = 5$ and $k = 3$, the sequence $SSFSF$ will occur with probability $p \times p \times (1 - p) \times p \times (1 - p) = p^3(1 - p)^2$. The total number of possible sequences with k successes and $n - k$ failures is $\binom{n}{k}$, since the binomial coefficient $\binom{n}{k}$ is the total number of ways to choose k different positions from n available positions, see Section 1.1.

The expected value and standard deviation of X are given by

$$\boxed{E(X) = np \ \text{ and } \ \sigma(X) = \sqrt{np(1 - p)}.}$$

The simplest way to derive these formulas is to use indicator variables. Write X as $X = I_1 + \cdots + I_n$, where I_k is 1 if the kth trial is a success and I_k is 0 otherwise. The Bernoulli variable I_k has $E(I_k) = p$ and $\sigma(I_k) = \sqrt{p(1 - p)}$, see Problem 2.45. Applying the linearity property for expectation and the $\sqrt{n}$-law for the standard deviation (the I_k's are independent of each other), you get

$$E\left(\sum_{k=1}^{n} I_k\right) = \sum_{k=1}^{n} E(I_k) = np \ \text{ and } \ \sigma\left(\sum_{k=1}^{n} I_k\right) = \sqrt{p(1 - p)} \times \sqrt{n}.$$

The binomial distribution is a versatile probability distribution and has numerous applications.

Example 3.1. A military early-warning installation is constructed in a desert. The installation consists of seven detectors including two reserve detectors. If fewer than five detectors are working, the installation ceases to function. Every six months an inspection of the installation is mounted, and at that time all detectors are replaced by new ones. There is a probability of 0.05 that any given detector will fail between two inspections. The detectors are all in operation and act independently of one another. What is the probability that the system will cease to function between two inspections?

Solution. Let the random variable X denote the number of detectors that will cease to function between two inspections. The random variable X has a binomial distribution with parameters $n = 7$ and $p = 0.05$. The probability that the system will cease to function between two inspections is

$$P(X > 2) = \sum_{k=3}^{7} \binom{7}{k} 0.05^k \, 0.95^{7-k} = 0.0038.$$

The binomial distribution can be used to solve the famous *problem of points*. In 1654, this problem was posed to Pascal and Fermat by the compulsory gambler Chevalier de Méré. Mathematics historians believe that the Chevalier posed the following problem: "Two players play a chance game of three points and each player has staked 32 pistoles. How should the sum be divided if they break off prematurely when one player has two points and the other player has one point?" A similar problem was earlier posed by the Italian mathematician Luca Pacioli in 1494 and led to heated discussions among Italian mathematicians in the 16th century, but none of them could come up with a satisfactory answer. The starting insight for Pascal and Fermat was that what is important is not so much the number of points each player has won yet, but the ultimate win probabilities of the players if the game were to continue at the point of stopping. The stakes should be divided in proportion to these win probabilities. Today the solution to the problem is obvious, but it was not at all obvious how to solve the problem in a time that the theory of probability was at an embryonic stage. The next example analyzes the problem of points in a modern outfit.

Example 3.2. In the World Series Baseball, the final two teams play a series consisting of no more than seven games until one of the teams has won four games. The winner takes all of the prize money of $1 000 000. In one such a final, two teams are pitted against one another and the stronger team will win any given game with a probability of 0.55. Unexpectedly, the competition must be suspended when the weaker team leads two games to one. How should the prize money be divided if the remaining games cannot be played?

Solution. At the point of stopping, the weaker team is 2 points away from the required 4 points and the stronger team 3 points. In the actual game, at most $2 + 3 - 1 = 4$ more matches would be needed to declare a winner. A trick to solve the problem is to imagine that four additional matches would be played. The probability of the weaker team being the ultimate winner if the original game was to be continued is the same as the probability that the weaker team would win two or more matches in four additional matches (explain!). The latter probability is equal to the binomial probability

$$\sum_{k=2}^{4} \binom{4}{k} 0.45^k \, 0.55^{4-k} = 0.609019.$$

The weaker team should receive $609 019 and the stronger team $390 981.

A less famous but still interesting problem from the history of probability is the Newton-Pepys problem. Isaac Newton was not much interested in probability. Nevertheless, Newton solved the following dice problem brought to him by Samuel Pepys who was a president of the Royal Society of London and apparently a gambling man. Which game is more likely to win: at least one six in one throw of six dice, at least two sixes in one throw of twelve dice, or at least three sixes in one throw of eighteen dice? What do you think? Pepys believed that the last option was the most favorable one.

Problem 3.1. A fair coin is to be tossed six times. You win two dollars if heads appears exactly three times (the expected number) and you lose one dollar otherwise. Is this game advantageous to you? (answer: no, your win probability is $\frac{5}{16}$)

Problem 3.2. Each day, the teacher randomly draws the names of four pupils in a class of 25 pupils. The homework of those four pupils is checked. What is the probability that your name will be drawn three or more times in the next five days? (answer: 0.0318)

Problem 3.3. Daily Airlines flies from Amsterdam to London every day. The plane has a passenger capacity of 150. The airline management has made it a policy to sell 160 tickets for this flight in order to protect themselves against no-show passengers. Experience has shown that the probability of a passenger being a no-show is equal to 0.08. The booked passengers act independently of each other. What is the probability that some passengers will have to be bumped from the flight? (answer: 0.0285)

Problem 3.4. Chuck-a-Luck is a carnival game of chance. To play this game, the player chooses one number from the numbers $1, \ldots, 6$.

Then three dice are rolled. If the player's number does not come up at all, the player loses 10 dollars. If the chosen number comes up one, two, or three times, the player wins $10, $20, or $30 respectively. What are the expected value and the standard deviation of the win for the house per wager? (answer: $0.787 and $11.13)

Problem 3.5. Suppose that 500 debit cards are stolen in a certain area. A thief can make three attempts to guess the four-digit pin code. The debit card is blocked after three unsuccessful attempts. What is the probability that the pin code of two or more debit cards is guessed correctly, assuming that each four-digit pin code is equally likely? (answer: 0.01017)

Problem 3.6. In the final of the World Series Baseball, two unevenly matched teams play a series consisting of at most seven games until one of the two teams has won four games. The probability that the weaker team will win any given game is 0.45, and the outcomes of the games are independent. What is the probability of the weaker team winning the final? (answer: 0.3917) What is the probability of the weaker team winning the final after six games? (answer: 0.1240)

Problem 3.7. What is the expected number of values showing up two or more times when six fair dice are rolled? (answer: 1.579) *Hint*: use indicator variables and the linearity of expectation.

Problem 3.8. In an ESP-experiment a medium has to guess the correct symbol on each of 250 Zener cards. Each card has one of the five possible Zener symbols on it and each of the symbols is equally likely to appear. The medium will get $100 000 dollars if he gives 82 or more correct answers. What is the probability that the medium must be paid out? (answer: 1.36×10^{-6})

Problem 3.9. You enter a gambling house (stock market) with a bankroll of $100, and you are going to play a game with 10 sequential bets. Each time, you bet your whole current bankroll. A fair coin is tossed. Your current bankroll increases with 70% if heads appears and decreases with 50% if tails appears (an expected return of 10%

for each bet!). What is the probability that your starting bankroll will be more than halved after 10 bets? (answer: 0.6230)

Problem 3.10. Seven friends are having a pleasant evening at the pub. Eventually they decide to play a coin game to determine how the beer will be paid for. Each of them tosses a fair coin and those who have tossed heads toss their coins again. This continues until there is one person left who has tossed heads or until there is no one left. In the first situation, that person pays for the beer; in the other situation, the friends share the bill. What is the probability that one unlucky person will have to pay the bill? (answer: 0.7211) *Hint*: use a recursion for the general case of n friends.

3.2 The hypergeometric distribution

The hypergeometric distribution is a discrete distribution that is closely related to the binomial distribution. The difference is that the trials in the hypergeometric context are not independent. The hypergeometric distribution describes the probability distribution of the number of successes when sampling *without replacement* from a finite population consisting of elements of two kinds. Think of an urn containing red and white balls or a shipment containing good and defect items. Suppose the population has R elements of the first type (for convenience, called successes) and W elements of the second type (called failures). Let n be the given number of elements that are randomly drawn from the population without replacement. Denote by the random variable X the number of successful elements drawn. Then, X has the *hypergeometric probability distribution*

$$P(X = r) = \frac{\binom{R}{r}\binom{W}{n-r}}{\binom{R+W}{n}} \quad \text{for } r = 0, 1, \ldots, R,$$

with the convention that $\binom{a}{b} = 0$ for $b > a$. The derivation of this distribution was already given in Section 1.1 and went as follows. Imagine that the elements are labeled so that all elements are distinguishable. Then the total number of possible combinations of n distinguishable elements is $\binom{R+W}{n}$ and among those combinations there are $\binom{R}{r} \times \binom{W}{n-r}$ combinations with r elements of the first type and $n - r$ elements of the second type. Each combination is equally likely and so $P(X = r)$ is the ratio of $\binom{R}{r} \times \binom{W}{n-r}$ and $\binom{R+W}{n}$.

The expected value and the standard deviation of X are

$$E(X) = np \quad \text{and} \quad \sigma(X) = \sqrt{np(1-p)\frac{R+W-n}{R+W-1}},$$

where $p = \frac{R}{R+W}$. A probabilistic derivation of the expected value and the variance will be given as a bonus at the end of this section. The hypergeometric distribution can be approximated by a binomial distribution with parameters n and $p = \frac{R}{R+W}$ when n is much smaller than $R + W$.

The best example for the hypergeometric distribution is the lottery, see Section 1.1. The hypergeometric model has many applications and shows up in various disguises, which at first sight have little to do with the classical model of red and white balls in an urn.

Example 3.3. In a close election between two candidates A and B in a small town, the winning margin of candidate A is 1 422 to 1 405 votes. However, 101 votes are found to be illegal and have to be thrown out. It is not said how the illegal votes are divided between the two candidates. Assuming that the illegal votes are not biased in any particular way and the count is otherwise reliable, what is the probability that the removal of the illegal votes will change the result of the election?

Solution. The problem can be translated into the urn model with 1 422 red and 1 405 white balls. If a is the number of illegal votes for candidate A and b the number of illegal votes for candidate B, then candidate A will no longer have most of the votes only if $a - b \geq 17$. Since $a + b = 101$, the inequality $a - b \geq 17$ boils down to $2a \geq 101 + 17$, or $a \geq 59$. The probability that the removal of the illegal votes will change the election result is the same as the probability of getting 59 or more red balls when randomly picking 101 balls from an urn with 1 422 red and 1 405 white balls. This probability is

$$\sum_{a=59}^{101} \frac{\binom{1\,422}{a}\binom{1\,405}{101-a}}{\binom{2\,827}{101}} = 0.0592.$$

The following example deals with a problem known as the German Tank Problem, in which statisticians helped the Allies estimate the number of tanks the Germans produced in World War II.

Example 3.4. In a bag there are n balls numbered as $1, \ldots, n$, where n is unknown. You can win a prize by guessing the right number of balls in the bag. To help you make a sensible guess, you are told that four balls drawn at random from the bag without replacement have the numbers 26, 33, 106, 108. How many balls should you estimate are in the bag?

Solution. Suppose that r numbers are drawn at random from 1 to n without replacement. Let the random variable M be the largest number drawn. The hypergeometric model is used to get $P(M = m)$. Fix m and imagine that the numbers 1 to $m - 1$ are green numbers, number m is a blue number, and the other numbers are red numbers. Then $P(M = m)$ can be interpreted as the probability of getting $r - 1$ green numbers and one blue number when drawing r numbers at random from 1 to n without replacement. This gives

$$P(M = m) = \frac{\binom{m-1}{r-1}\binom{1}{1}}{\binom{n}{r}} \quad \text{for } m = r, \ldots, n.$$

Next, using the binomial identity $\sum_{m=r}^{n} \binom{m}{r} = \binom{n+1}{r+1}$, you find

$$E(M) = \sum_{m=r}^{n} m \frac{\binom{m-1}{r-1}}{\binom{n}{r}} = \frac{r}{\binom{n}{r}} \sum_{m=r}^{n} \binom{m}{r} = \frac{r\binom{n+1}{r+1}}{\binom{n}{r}} = \frac{r(n+1)}{r+1}.$$

Solving n in terms of r and $E(M)$ gives

$$n = E(M)\left(1 + \frac{1}{r}\right) - 1.$$

Suppose the observed value of the largest of the r numbers drawn is m^*. Taking m^* as the best guess for $E(M)$, you estimate the unknown n by

$$n \approx m^*\left(1 + \frac{1}{r}\right) - 1.$$

Substituting $r = 4$ and $m^* = 108$ in this expression, you get the estimate $108(1 + 0.25) - 1 = 134$ for the number of balls in the bag.

Problem 3.11. In the game "Lucky 10" twenty numbers are drawn from the numbers 1 to 80. You tick 10 numbers on the game form. What is the probability of matching 5 or more of the 20 numbers drawn? (answer: 0.0647)

Problem 3.12. A bowl contains 10 red and 15 white balls. You randomly pick without replacement one ball at a time until you have 5 red balls. What the probability that more than 10 picks are needed? (answer: 0.6626)

Problem 3.13. For a final exam, your professor gives you a list of 15 items to study. He indicates that he will choose eight for the actual exam. You will be required to answer correctly at least five of those. You decide to study 10 of the 15 items. What is the probability that you will pass the exam? (answer: $\frac{9}{11}$)

Problem 3.14. What is the probability that a bridge player has more than one ace among 13 randomly dealt cards from a standard deck of 52 cards given that the player has an ace? (answer: 0.3696) How does this probability change if the player had the ace of hearts? (answer: 0.5612)

Problem 3.15. Two people, perfect strangers to one another, both living in the same city of one million inhabitants, meet each other. Each has approximately 500 acquaintances in the city. Assuming that for each of the two people, the acquaintances represent a random sampling of the city's various population sectors, what is the probability of the two people having an acquaintance in common?

Probabilistic derivation of the expected value and variance

It is instructive to give a probabilistic derivation of the expected value and the variance of the hypergeometric distribution. Take an urn with R red balls and W white balls. There is no restriction to assume that the $R + W$ balls are distinguishable. Imagine that the $R + W$ balls are randomly ordered in a row and that the balls in the first n positions represent the n balls that are randomly drawn from the urn without replacement. Using the assumption of distinguishable balls, the total number of orderings is $(R + W)!$ and the number of orderings for which there is a red ball at a given position i is $R(R+W-1)!$. Then, letting the indicator variable X_i be 1 if there is a red ball at the ith position and be 0 otherwise,

$$P(X_i = 1) = \frac{R(R+W-1)!}{(R+W)!} = \frac{R}{R+W} \quad \text{for } i = 1, \ldots, n.$$

The number of orderings for which there is a red ball both at the ith position and at the jth position is $R(R-1)(R+W-2)!$ and so

$$P(X_i = X_j = 1) = \frac{R(R-1)(R+W-2)!}{(R+W)!} = \frac{R(R-1)}{(R+W)(R+W-1)}.$$

Let the random variable X be the number of red balls among the balls in the first n positions, then $X = \sum_{i=1}^{n} X_i$. Since $E(X_i) = P(X_i = 1)$, you get by the linearity of expectation that $E(X) = n\frac{R}{R+W}$. By the dependence of the X_i, the formula

$$E(X^2) = E[(X_1 + \ldots + X_n)^2] = \sum_{i=1}^{n} E(X_i^2) + 2\sum_{i=1}^{n-1}\sum_{j=i+1}^{n} E(X_i X_j)$$

must be used to calculate $\text{var}(X)$. Since $E(X_i^2) = P(X_i = 1)$ for all i and $E(X_i X_j) = P(X_i = 1 \text{ and } X_j = 1)$ for all $i \neq j$, you find after a little algebra that

$$E(X^2) = n\frac{R}{R+W} + n(n-1)\frac{R(R-1)}{(R+W)(R+W-1)}.$$

Next, the formula for $\text{var}(X)$ follows by using $\text{var}(X) = E(X^2) - E^2(X)$.

3.3 The Poisson distribution

A random variable X with $0, 1, \ldots$ as possible values is said to have a *Poisson distribution* with parameter $\lambda > 0$ if

$$P(X = k) = e^{-\lambda}\frac{\lambda^k}{k!} \quad \text{for } k = 0, 1, \ldots,$$

where $e = 2.71828\ldots$ is the Euler number. The expected value and the standard deviation of X are

$$E(X) = \lambda \quad \text{and} \quad \sigma(X) = \sqrt{\lambda}.$$

The proof is simple. Noting that $k! = k \times (k-1)!$ and $\sum_{j=0}^{\infty} e^{-\lambda} \frac{\lambda^j}{j!} = 1$, you get

$$E(X) = \sum_{k=0}^{\infty} k e^{-\lambda} \frac{\lambda^k}{k!} = \lambda \sum_{k=1}^{\infty} e^{-\lambda} \frac{\lambda^{k-1}}{(k-1)!} = \lambda \sum_{j=0}^{\infty} e^{-\lambda} \frac{\lambda^j}{j!} = \lambda.$$

In the same way, you get $E[X(X-1)] = \lambda^2$, which gives $E(X^2) = \lambda^2 + E(X)$ and so $\sigma^2(X) = \lambda^2 + \lambda - \lambda^2 = \lambda$.

The Poisson distribution is useful in studying rare events. It is a very good approximation to the binomial distribution of the total number of successes in a *very large* number of independent trials each having a *very small* probability of success. That is, the binomial probability $\binom{n}{k} p^k (1-p)^{n-k}$ of getting k successes in n independent trials each having success probability p tends to $e^{-\lambda} \frac{\lambda^k}{k!}$ for all k if $n \to \infty$ and $p \to 0$ such that $np \to \lambda$ for a constant $\lambda > 0$. To avoid technicalities, we prove this only for the case that np is kept fixed on the value λ ($p = \frac{\lambda}{n}$). The proof then goes as follows: write the binomial probability $\binom{n}{k} p^k (1-p)^{n-k}$ as

$$\binom{n}{k} \left(\frac{\lambda}{n}\right)^k \left(1 - \frac{\lambda}{n}\right)^{n-k} = \frac{n!}{k!\,(n-k)!} \frac{\lambda^k}{n^k} \frac{(1-\lambda/n)^n}{(1-\lambda/n)^k}$$

$$= \frac{\lambda^k}{k!} \left(1 - \frac{\lambda}{n}\right)^n \left[\frac{n!}{n^k\,(n-k)!}\right] \left(1 - \frac{\lambda}{n}\right)^{-k}.$$

The first term $\frac{\lambda^k}{k!}$ does not depend on n. The second term $\left(1 - \frac{\lambda}{n}\right)^n$ tends to $e^{-\lambda}$ if n tends to infinity. The third term $\frac{n!}{n^k(n-k)!}$ can be written as: $n(n-1)\cdots(n-k+1)/n^k$ and thus equals $(1 - \frac{1}{n})(1 - \frac{2}{n})\cdots(1 - \frac{k-1}{n})$. Thus, for fixed k, the third term tends to 1 if $n \to \infty$. For fixed k, the last term $(1 - \frac{\lambda}{n})^{-k}$ also tends to 1 if $n \to \infty$. This completes the proof.

A very important observation is that only the product value $\lambda = np$ is relevant for the Poisson approximation to the binomial distribution with parameters n and p. You do not need to know the particular values of the number of trials and the success probability. It is enough to know what the expected (or average) value of the total number of successes is. This is an extremely useful property when it comes to practical applications. The physical background of the Poisson distribution, as a distribution of the total number of successes in a large number of trials each having a small probability of success, explains why this distribution has so many practical applications: the annual number of damage claims at insurance companies, the annual number of severe traffic accidents in a given region, the annual number of stolen credit cards, the annual number of fatal shark bites worldwide, etc.[16] Also, the Poisson distribution often provides a good description of many situations involving points randomly distributed in a bounded region in the plane or space.

The mathematical derivation of the Poisson distribution assumes that the trials are physically independent of each other, but, in many practical situations, the Poisson distribution also appears to give good approximations when there is a 'weak' dependence between the outcomes of the trials. The Poisson heuristic is especially useful for quickly arriving at good approximated results in problems for which it would otherwise be difficult to find exact solutions. Applications of the Poisson heuristic will be given in Chapter 4.

Example 3.5. There are 500 people present at a gathering. For the fun of it, the organizers have decided that all of those whose birthday is that day will receive a present. How many presents are needed to ensure a less than 1% probability of having too few presents?

Solution. Let the random variable X represent the number of individuals with a birthday on the day of the gathering. Leap year day, February 29, is discounted, and apart from that, it is assumed that every day of the year is equally likely as birthday. The distribution of X can then be modeled by a binomial distribution with parameters $n = 500$ and $p = \frac{1}{365}$. Calculating $P(X > k)$ reveals that

$$P(X > 4) = 0.0130 \quad \text{and} \quad P(X > 5) = 0.0028,$$

and so five presents suffice. Since $n = 500$ is large and $p = \frac{1}{365}$ is small, the binomial distribution of X can be approximated by a

---

[16]Under rather weak conditions the Poisson distribution also applies under non-identical success probabilities of the trials.

Table 1: Binomial and Poisson probabilities

k	0	1	2	3	4	5	6
bin	0.2537	0.3484	0.2388	0.1089	0.0372	0.0101	0.0023
Poi	0.2541	0.3481	0.2385	0.1089	0.0373	0.0102	0.0023

Poisson distribution with expected value $\lambda = np = \frac{500}{365}$. For comparison, for both the binomial and the Poisson distribution, Table 1 gives the probability that exactly k persons have a birthday on the day of the gathering for $k = 0, 1, \ldots, 6$. The probabilities agree very well.

The z-score test

A practically useful characteristic of the Poisson distribution is that the probability of a value more than three standard deviations removed from the expected value is very small (10^{-3} or smaller) when the expected value λ is sufficiently large. A rule of thumb is $\lambda \geq 25$. This rule is very useful for judging the value of all sorts of statistical facts reported in the media. In order to judge how exceptional a certain random outcome is, you measure how many standard deviations the outcome is removed from the expected value. This is called the *z-score test* in statistics. For example, suppose that in a given year the number of break-ins occurring in a given area increases more than 15% from an average of 64 break-ins per year to 75 break-ins. Since the z-score is $(75 - 11)/\sqrt{64} = 1.38$, the increase can be explained as a chance fluctuation and so there is no reason to demand the resignation of the police officer.

Example 3.6. The Pegasus Insurance Company has introduced a policy that covers certain forms of personal injury with a standard payment of \$100 000. On average, 100 claims per year lead to payment. There are many tens of thousands of policyholders. What can be said about the probability that more than 15 million dollars will have to be paid out in the space of a year?

Solution. In fact, every policyholder conducts a personal experiment in probability after purchasing this policy, which can be considered to be "successful" if the policyholder files a rightful claim during the ensuing year. In view of the many policyholders, there is a large number of independent probability experiments each having a very small probability of success. Therefore, the Poisson model can be used. Denoting by the random variable X the total number of claims that will be approved for payment during the year of coverage, the random variable X can be modeled by a Poisson distribution with parameter $\lambda = 100$. The probability of having to pay out more than 15 million dollars is given by $P(X > 150)$. Since $E(X) = 100$ and $\sigma(X) = 10$, a value of 150 claims lies five standard deviations above the expected value. Thus, without doing any further calculations, you can draw the conclusion that the probability of paying out more than 15 million dollars in the space of a year must be extremely small. The precise value of the probability is 1.23×10^{-6}. Not a probability the insurance executives need to worry about.

For a binomially distributed random variable X with parameters n and p, it is also true that almost all probability mass from the distribution lies within three standard deviations from the expected value when $np(1 - p)$ is sufficiently large. A rule of thumb for this is $np(1 - p) > 20$. A beer brewery once made brilliant use of this. In a television advertisement spot broadcast during the American Super Bowl final, 100 beer drinkers were asked to do a blind taste test comparing beer brewed by the sponsored brewery, and beer brewed by a competitor. The brilliance of the stunt is that the 100 beer drinkers invited to participate were regular drinkers of the brand made by the competitor. In those days, all brands of American beer tasted more or less the same, and most drinkers weren't able to distinguish between brands. The marketers of the sponsored beer could therefore be pretty sure that more than 35% of the participants in the stunt would prefer the sponsored beer over their regular beer. The target value of 35 is $(50 - 35)/5 = 3$ standard deviations below the expected value of 50 and so the binomial probability of falling below 35 is very small. This did, in fact, occur, and made quite an impression on the television audience.

As another illustration of the z-score test, in Problem 3.8 the z-value for 82 or more good answers is $(82 - 50)/\sqrt{40} \approx 5.06$ and so you can conclude without any further calculations that the probability of payout is extremely small.

Poissonization of the binomial experiment

In the experiment with a fixed number of independent Bernoulli trials each having the same probability of success, the number of successes and the number of failures both have a binomial distribution. These two distributions are dependent. The picture changes in a surprising way if you randomize the number of trials according to a Poisson distribution.

The experiment is to have a sequence of independent Bernoulli trials each having the same probability p of success, where the number of trials is Poisson distributed with parameter λ. Then the following result holds:

(a) *The number of successes and the number of failures are Poisson distributed with parameters λp and $\lambda(1 - p)$.*
(b) *The number of successes and the number of failures are independent of each other.*

The proof goes as follows. Let the random variable X be the number of successes and the random variable Y be the number of failures. The probability $P(X = j$ and $Y = k)$ can be written as $P(A$ and $B)$, where A is the event that the number of trials is $j + k$ and B is the event that there are exactly j successes among the $j + k$ Bernoulli trials. By $P(A$ and $B) = P(A)P(B \mid A)$,

$$P(X = j \text{ and } Y = k) = e^{-\lambda}\frac{\lambda^{j+k}}{(j+k)!}\binom{j+k}{j}p^j(1-p)^k \text{ for all } j, k.$$

Writing $e^{-\lambda} = e^{-\lambda p}e^{-\lambda(1-p)}$ and $\binom{j+k}{j} = \frac{(j+k)!}{j!k!}$, it follows that

$$P(X = j \text{ and } Y = k) = e^{-\lambda p}\frac{(\lambda p)^j}{j!} \times e^{-\lambda(1-p)}\frac{(\lambda(1 - p))^k}{k!} \text{ for all } j, k.$$

Summing $P(X = j$ and $Y = k)$ over k gives $P(X = j) = e^{-\lambda p}(\lambda p)^j/j!$ for all j. Similarly, $P(Y = k) = e^{-\lambda(1-p)}(\lambda(1-p))^k/k!$ for all k. Since $P(X = j$ and $Y = k) = P(X = j)P(Y = k)$ for all j, k, the Poisson distributed random variables X and Y are independent. A surprising finding!

The result for the binomial experiment can be straightforwardly extended to the multinomial experiment, see Problem 3.24. In this experiment, each trial has r different possible outcomes with probabilities $p_1, \ldots, p_r$. Think of putting balls one at a time into one of r bins, where each ball is put into the ith bin with a given probability p_i. In the multinomial experiment with n independent trials, the calculation of probabilities of interest can become computationally quite demanding. However, tractable results are obtained when the number of trials is randomized and has a Poisson distribution with an expected value of $\lambda = n$. For large n, this Poissonized model can be used as an approximation to the model with a fixed number of n trials.

Problem 3.16. What is the Poisson approximation for the sought probability in Problem 3.5? (answer: 0.01019)

Problem 3.17. What is the probability of the jackpot falling in lotto 6/42 when 5 million tickets are randomly filled in? (answer: 0.6145)

Problem 3.18. In a coastal area, the average number of serious hurricanes is 3.1 per year. Use an appropriate probability model to calculate the probability of a total of more than 5 serious hurricanes in the next year (answer: 0.0943)

Problem 3.19. The low earth orbit contains many pieces of space debris. It is estimated that an orbiting space station will be hit by space debris beyond a critical size and speed on average once in 400 years. Estimate the probability that a newly launched space station will not be penetrated in the first 20 years. (answer: 0.951)

Problem 3.20. Suppose r dice are simultaneously rolled each time. A roll in which each of the r dice shows up a six is a called a king's roll (generalization of de Méré's dice problem). For larger values of r, what is the probability of not getting a king's roll in $4 \times 6^{r-1}$ rolls of the r dice? (answer: $e^{-2/3} = 0.5134$)

Problem 3.21. In a particular rural area, postal carriers are attacked by dogs 324 times per year on average. Last year there were 379 attacks. Is this exceptional? (answer: yes, the z-score is 3.1)

Problem 3.22. An average of 20 fires occur in a given region each year. Last year, the number of fires increased by 35% over this average. Is this exceptional? (answer: no, the z-score is 1.57)

Problem 3.23. On average there are 4.2 fatal shark attacks each year worldwide. What is the probability that there will more than seven fatal shark attacks next year worldwide? (answer: 0.0639)

Problem 3.24. You put balls, one at a time, into one of b bins labeled as 1 to b. Any ball is put into bin j with probability p_j. The number of balls to be put into the bins is Poisson distributed with an expected value of λ. Let X_j be the number of balls that will be put into bin j. Verify that X_j is Poisson distributed with an expected value of λp_j for $j = 1, \ldots, b$. Show that $X_1, \ldots, X_b$ are independent of each other.

3.4 The normal probability density

Many probabilistic situations are better described by a continuous random variable rather than a discrete random variable. Think of the annual rainfall in a certain area or the decay time of a radioactive particle. Calculations in probability and statistics are often greatly simplified by approximating the probability mass function of a discrete random variable by a continuous curve. As illustrated in Figure 4, the probability mass function of the binomial distribution with parameters n and p can very well be approximated by the graph of the continuous function

$$f(x) = \frac{1}{\sigma\sqrt{2\pi}} e^{-\frac{1}{2}(x-\mu)^2/\sigma^2}$$

with $\mu = np$ and $\sigma = \sqrt{np(1-p)}$ if n is sufficiently large, say $np(1-p) \geq 20$. This function $f(x)$ is called the *normal density function* (or the *Gaussian density function*). Likewise, the probability mass function of the Poisson distribution with parameter λ can be approximated by the graph of the normal density function $f(x)$ with $\mu = \lambda$ and $\sigma = \sqrt{\lambda}$ if λ is sufficiently large, say $\lambda \geq 25$. In Figure 4, the parameters of the binomial and the Poisson distributions are $(n = 125, p = \frac{1}{5})$ and $\lambda = 25$.

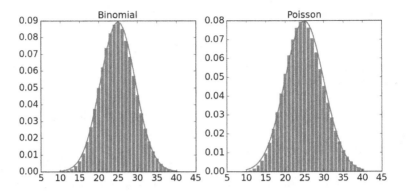

Figure 4: Normal approximation.

You have now arrived at the normal distribution, which is the most
important probability distribution. A continuous random variable X
is said to have a *normal distribution* with parameters μ and $\sigma > 0$ if

$$P(X \leq x) = \frac{1}{\sigma\sqrt{2\pi}} \int_{-\infty}^{x} e^{-\frac{1}{2}(t-\mu)^2/\sigma^2} dt \quad \text{for } -\infty < x < \infty.$$

The notation $N(\mu, \sigma^2)$ is often used for a normally distributed random variable X with parameters μ and σ.

The normal density function

$$f(x) = \frac{1}{\sigma\sqrt{2\pi}} e^{-\frac{1}{2}(x-\mu)^2/\sigma^2}$$

is found as the derivative of the probability distribution function
$F(x) = P(X \leq x)$. The non-negative function $f(x)$ can be shown to
integrate to 1 over $(-\infty, \infty)$. The parameters μ and σ of the normal
density are the expected value and the standard deviation of X,

$$\mu = \int_{-\infty}^{\infty} xf(x)\,dx \quad \text{and} \quad \sigma^2 = \int_{-\infty}^{\infty} (x-\mu)^2 f(x)\,dx.$$

The derivation of these formulas is beyond the scope of this book.
The following remarks are made. The normal density function $f(x)$

is maximal for $x = \mu$ and is symmetric around the point $x = \mu$. The point $x = \mu$ is also the median of the normal probability distribution.[17] About 68.3% of the probability mass of a normally distributed random variable with expected value μ and standard deviation σ is between the points $\mu - \sigma$ and $\mu + \sigma$, about 95.4% between $\mu - 2\sigma$ and $\mu + 2\sigma$, and about 99.7% between $\mu - 3\sigma$ and $\mu + 3\sigma$. These facts are displayed in Figure 5 and will be explained below after having introduced the standard normal distribution.

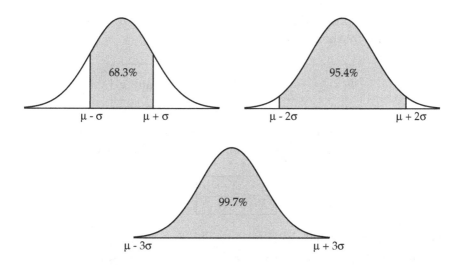

Figure 5: Characteristics of the normal density function.

The normal distribution has the important characteristic that $aX + bY$ is normally distributed for any constants a and b if the random variables X and Y are normally distributed and independent of each other. The expected value μ and the standard deviation σ of $aX + bY$ are then equal to

$$\mu = aE(X) + bE(Y) \quad \text{and} \quad \sigma = \sqrt{a^2\sigma^2(X) + b^2\sigma^2(Y)}.$$

---

[17]The median of a continuous random variable is defined as a point such that the random variable has 50% of its probability mass left from that point and 50% of its probability mass right from that point. It is noted that for a continuous random variable each individual point has probability mass zero.

In particular, the random variable $aX + b$ is $N(a\mu + b, a^2\sigma^2)$ distributed if the random variable X is $N(\mu, \sigma^2)$ distributed.

The *standard normal distribution* is the normal distribution with expected value 0 and standard deviation 1. This distribution is usually denoted as the $N(0, 1)$ distribution. For a standard normally distributed random variable Z, the notation

$$\Phi(z) = P(Z \leq z)$$

is used for the probability distribution function of Z. The function $\Phi(z)$ is given by the famous integral

$$\Phi(z) = \frac{1}{\sqrt{2\pi}} \int_{-\infty}^{z} e^{-\frac{1}{2}x^2}\, dx \quad \text{for any } z.$$

If X has an $N(\mu, \sigma^2)$ distribution, then, by $E(aX + b) = aE(X) + b$ and $\sigma^2(aX + b) = a^2\sigma^2(X)$, the *normalized* random variable

$$Z = \frac{X - \mu}{\sigma}$$

has a standard normal distribution. This is a very useful result for the calculation of the probabilities $P(X \leq x)$. Writing

$$P(X \leq x) = P\left(\frac{X - \mu}{\sigma} \leq \frac{x - \mu}{\sigma}\right),$$

you see that $P(X \leq x)$ can be calculated as

$$P(X \leq x) = \Phi\left(\frac{x - \mu}{\sigma}\right).$$

In particular, you have $P(X \leq \mu) = \Phi(0) = 0.5$. Using the formula $P(a < X \leq b) = P(X \leq b) - P(X \leq a)$, it follows that $P(a < X \leq b)$ can be calculated as

$$P(a < X \leq b) = \Phi\left(\frac{b - \mu}{\sigma}\right) - \Phi\left(\frac{a - \mu}{\sigma}\right) \quad \text{for any } a < b.$$

This result explains the percentages in Figure 5. As an example, an $N(\mu, \sigma^2)$ distributed random variable has about 95.4% of its probability mass between $\mu - 2\sigma$ and $\mu + 2\sigma$, since $\Phi(2) - \Phi(-2) = 0.9545$.

As an illustration of the normal distribution, the length of Northern European boys who are born after a gestational period between 38 and 42 weeks has a normal distribution with an expected value of 50.9 cm and a standard deviation of 2.4 cm at birth. How exceptional is it that a boy at birth has a length of 48 cm? A quick answer to this question can be given by using the z-score test: a length of 48 cm is $\frac{50.9 - 48}{2.4} = 1.2083$ standard deviations below the expected value and this is not exceptional. The undershoot probability $\Phi(-1.2083) = 0.1135$ corresponds to a z-score of -1.2083.

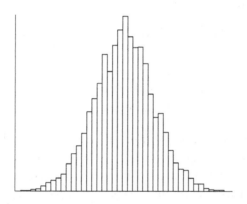

Figure 6: Histogram of height measurements.

As said before, the normal distribution is the most important continuous distribution. Many stochastic situations in practice can be modeled with the help of the normal distribution. For example, the annual rainfall in a certain area, the cholesterol level of an adult male of a specific racial group, errors in physical measurements, the length of men in a certain age group, etc. Figure 6 displays a histogram of height measurements of a large number of men in a certain age group. A histogram divides the range of values covered by the measurements into intervals of the same width, and shows the proportion of the measurements in each interval. You see that the histogram has the characteristic bell-shaped form of the graph of the normal density function. Making the width of the intervals smaller and smaller and the number of observations larger and larger, the graph of the histogram changes into the graph of a normal density function.

Problem 3.25. The annual rainfall in Amsterdam has a normal distribution with an expected value of 799.5 mm and a standard deviation of 121.4 mm. What is the probability of having more than 1000 mm rainfall in Amsterdam next year? (answer: 0.0493)

Problem 3.26. Gestation periods of humans have a normal distribution with an expected value of 280 days and a standard deviation of 10 days. What is the percentage of births that are more than 15 days overdue? (answer: 0.0668)

Problem 3.27. The diameter of a 1 euro coin has a normal distribution with an expected value of 23.25 mm and a standard deviation of 0.10 mm. A vending machine accepts only 1 euro coins with a diameter between 22.90 mm and 23.60 mm. What is the probability that a 1 euro coin will not be accepted? (answer: 4.65×10^{-4})

Problem 3.28. The annual grain harvest in a certain area is normally distributed with an expected value of 15 000 tons and a standard deviation of 2 000 tons. In the past year the grain harvest was 21 500 tons. Is this exceptional? (answer: yes, the z-score is 3.075)

Problem 3.29. What is the standard deviation of the demand for a certain item if the demand has a normal distribution with an expected value of 100 and the probability of a demand exceeding 125 is 0.05? (answer: 15.2)

Problem 3.30. A stock return can be modeled by an $N(\mu, \sigma^2)$ distributed random variable. An investor believes that there is a 10% probability of a return below $80 and a 10% probability of a return above $120. What are the investor's estimates of μ and σ? (answer: $\mu = 100$ and $\sigma = 15.6$)

Problem 3.31. Verify that $P(|X - \mu| > k\sigma) = 2[1 - \Phi(k)]$ for any $k > 0$ if X is $N(\mu, \sigma^2)$ distributed.

3.5 Central limit theorem and the normal distribution

In this section, the most famous theorem of probability and statistics is discussed. This theorem explains why many stochastic situations in practice can be modeled with the help of a normal distribution. If a random variable can be seen as the result of the sum of a large number of small independent random effects, then it is approximately normally distributed. Mathematically, this result is expressed by the *central limit theorem*, which is the most celebrated theorem in probability and statistics. Loosely formulated,

> **if the random variables $X_1, X_2, \ldots, X_n$ are independent of each other and have each the same probability distribution with expected value μ and standard deviation σ, then the sum $X_1 + X_2 + \cdots + X_n$ has approximately a normal distribution with expected value $n\mu$ and standard deviation $\sigma\sqrt{n}$ if n is sufficiently large.**

Alternatively, it can be said: the sample mean $\frac{1}{n}(X_1 + X_2 + \cdots + X_n)$ has approximately a normal distribution with expected value μ and standard deviation $\frac{\sigma}{\sqrt{n}}$ if n is sufficiently large.[18] That is, for any x,

$$
\boxed{
\begin{aligned}
&P(X_1 + X_2 + \cdots + X_n \le x) \approx \Phi\left(\frac{x - n\mu}{\sigma\sqrt{n}}\right) \quad \text{for large } n \\
&P\left(\frac{X_1 + X_2 + \cdots + X_n}{n} \le x\right) \approx \Phi\left(\frac{x - \mu}{\sigma/\sqrt{n}}\right) \quad \text{for large } n.
\end{aligned}
}
$$

How large n should be depends on the shape of the probability distribution of the X_i; the more symmetric this distribution is, the sooner the normal approximation applies. To illustrate this, Figure 7 displays the probability histogram of the total sum obtained in n rolls of a symmetrical die for several values of n, and Figure 8 does the same thing for an asymmetrical die. The figures nicely show that the more *skewed* the probability mass function, the *larger* n must be so that $\sum_{k=1}^{n} X_k$ is approximately normally distributed.

---

[18]$\lim_{n \to \infty} P\left(\frac{X_1 + \cdots + X_n - n\mu}{\sigma\sqrt{n}} \le x\right) = \Phi(x)$ for all x is the mathematically precise formulation of the central limit theorem.

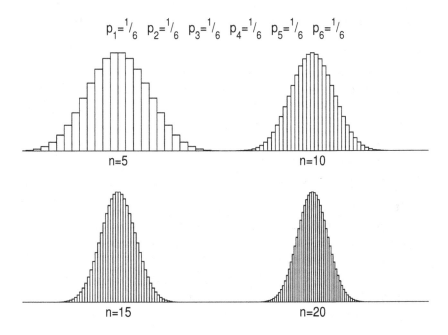

$p_1={}^1\!/_6 \quad p_2={}^1\!/_6 \quad p_3={}^1\!/_6 \quad p_4={}^1\!/_6 \quad p_5={}^1\!/_6 \quad p_6={}^1\!/_6$

n=5   n=10

n=15   n=20

Figure 7: Probability histograms when the die is symmetrical.

The central limit theorem explains why the histogram of the probability mass function of a binomially distributed random variable with parameters n and p can be nicely approximated by the graph of a normal density with expected valued np and standard deviation $\sqrt{np(1-p)}$ if n is sufficiently large: a binomial random variable can be written as the sum $X_1 + \cdots + X_n$ of n independent random variables X_i with $P(X_i = 1) = p$ and $P(X_i = 0) = 1 - p$.

The central limit theorem is extremely useful for both practical and theoretical purposes. To illustrate, consider Example 2.15 again. Using the fact that an $N(\alpha, \beta^2)$ distributed random variable has 95.4% of its probability mass between $\alpha - 2\beta$ and $\alpha + 2\beta$, there is a probability of about 95% that the net profit of the school after 500 bets will be between $250 - 2 \times 25 = 200$ dollars and $250 + 2 \times 25 = 300$ dollars.

The central limit theorem has an interesting history. The first version of this theorem was postulated in 1738 by the French-born English

p_1=0.2 p_2=0.1 p_3=0.0 p_4=0.0 p_5=0.3 p_6=0.4

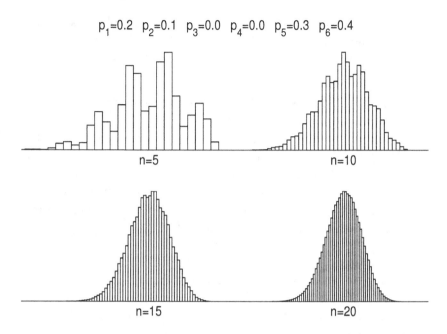

Figure 8: Probability histograms when the die is asymmetrical.

mathematician Abraham de Moivre, who used the normal distribution to approximate the distribution of the number of heads resulting from many tosses of a fair coin. De Moivre's finding was far ahead of its time, and was nearly forgotten until the famous French mathematician Pierre Simon Laplace rescued it from obscurity in his monumental work *Théorie Analytique des Probabilités*, which was published in 1812. Laplace expanded De Moivre's finding by approximating the binomial distribution with the normal distribution. But as with De Moivre, Laplace's finding received little attention in his own time. It was not until the nineteenth century was at an end that the importance of the central limit theorem was discerned when, in 1901, the Russian mathematician Aleksandr Lyapunov defined it in general terms and proved precisely how it worked mathematically.

The central limit theorem and the law of large numbers are two pillars of probability theory. The law of large numbers states that $\frac{1}{n}\sum_{k=1}^{n} X_k$ tends to $E(X)$ with probability one as n tends to infinity, and the central limit theorem enables you to give probabilistic error bounds on deviations of $\frac{1}{n}\sum_{k=1}^{n} X_k$ from $E(X)$ for large n. These matters will come back in Chapter 5 on simulation, including the statistical concept of confidence interval.

Example 3.7. For an expedition with a duration of one and a half years, a number of spare copies of a particular filter must be taken along. The filter will be used daily. The lifetime of the filter has a continuous probability distribution with an expected value of one week and a standard deviation of half a week. Upon failure a filter is replaced immediately by a new one. How many filters should be taken along with the expedition in order to ensure that there will be no shortage with a probability of at least 99%?

Solution. Suppose n filters are taken along for the expedition of 78 weeks. The probability of no shortage during the expedition is $P(X_1 + \cdots + X_n > 78)$, where X_i is the lifetime (in weeks) of the ith filter. The lifetimes of the filters are assumed to be independent of each other. Then, by the central limit theorem,

$$P(X_1 + \cdots + X_n > 78) = 1 - P(X_1 + \cdots + X_n \leq 78)$$
$$= 1 - P\left(\frac{X_1 + \cdots + X_n - n}{0.5\sqrt{n}} \leq \frac{78 - n}{0.5\sqrt{n}}\right) \approx 1 - \Phi\left(\frac{78 - n}{0.5\sqrt{n}}\right).$$

The requirement is that $1 - \Phi((78 - n)/(0.5\sqrt{n})) \geq 0.99$, and so you need the smallest value of n for which

$$\Phi\left(\frac{78 - n}{0.5\sqrt{n}}\right) \leq 0.01.$$

The solution of the equation $\Phi(x) = 0.01$ is $x = -2.326$ (the so-called 1% percentile[19]). Next, you solve the equation

$$\frac{78 - z}{0.5\sqrt{z}} = -2.326.$$

The solution is $z = 88.97$. Thus, 89 copies of the filter are needed.

---

[19]For any $0 < p < 1$, the $100p\%$ percentile z_p of the standard normal distribution is defined as the unique solution to $\Phi(x) = p$. For example, $z_{0.025} = -1.960$ and $z_{0.975} = 1.960$.

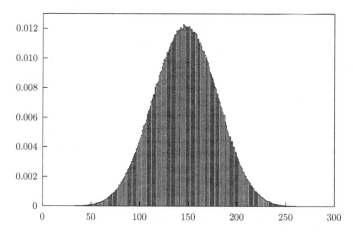

Figure 9: Probability histogram for $r = 6$ and $s = 49$.

Statistical application

The central limit theorem also applies to the following situation in which you have weakly dependent random variables. Suppose that r distinct numbers are sequentially drawn from the numbers 1 up to s, one at a time and at random. Let the random variables X_i represent the ith number drawn. For reasons of symmetry, each of the X_i has the same probability distribution, but they are not independent. If s is large and r is much smaller than s, the dependency between the X_i is weak. Then, it can be proved that $X_1 + \cdots + X_r$ is approximately $N(\mu, \sigma^2)$ distributed with

$$\mu = \frac{1}{2}r(s+1) \quad \text{and} \quad \sigma^2 = \frac{1}{12}r(s-r)(s+1).$$

This result can be used to demystify a widely advertised lottery system claimed to increase one's chances to win the lottery. As an example, let's take the lottery 6/49 in which six different numbers are randomly drawn from the numbers 1 up to 49. The lottery system is based on the bell curve and fools lottery players by suggesting them to choose six different numbers that add up to a number between 117 and 183. If you take $r = 6$ and $s = 49$ in the above probability model, you get $\mu = 150$ and $\sigma = 32.8$. The normal curve has about

68% of its probability mass between $\mu - \sigma$ and $\mu + \sigma$. That's why the lottery system suggests players to choose six lottery numbers that add up to a number between $150 - 33 = 117$ and $150 + 33 = 183$. It is true that the sum of the six winning numbers will fall between 117 and 183 with a probability of about 68%, but this lottery system does not increase the player's odds of winning a prize. The advice completely neglects the fact that there are many more combinations of six numbers whose sum falls in the middle of the sum's distribution than the combinations of six numbers whose sum falls in the tail of the distribution. Figure 9 displays the simulated frequency of the sum of the six winning lottery numbers, based on one million drawings. As you see, the probability histogram can accurately be approximated by a normal density function.

Problem 3.32. An insurance company has $20\,000$ policyholders. The amount claimed yearly by a policyholder has an expected value of $150 and a standard deviation of $750. Calculate an approximation for the probability that the total amount claimed in the coming year will exceed 3.3 million dollars. (answer: 0.0023)

Problem 3.33. Let the random variable H_n be the number of heads showing up in n tosses of a fair coin. What is the approximate distribution of $H_n - \frac{1}{2}n$ for large n? (answer: $N(0, (\frac{1}{2}\sqrt{n})^2)$)

Problem 3.34. In the random walk on the line, a drunkard takes each time a unit step to the right or to the left with equal probabilities, independently of the previous steps. Let D_n be the distance of the drunkard from the starting point after n steps. What is an approximation to $E(D_n)$ for n large? (answer: $\sqrt{2n/\pi}$)

Problem 3.35. The Nero Palace casino has a new, exciting gambling machine: the multiplying bandit. How does it work? The bandit has a lever or "arm" that the player can depress up to ten times. After each pull, an H (heads) or a T (tails) appears, each with probability $\frac{1}{2}$. The game is over as soon as heads appears or the player has pulled the arm ten times. The player wins $\$2^k$ if heads appears at the k pull for $1 \le k \le 10$, and wins $1500 otherwise. The stake for this game is $15. What are the expected value and the standard deviation of the casino's profit for each game? (answer: $3.5352 and $45.249) What is the approximate distribution of the casino's profit over 2 500 games? (answer: $N(8\,838, (2\,262)^2)$)

Problem 3.36. A new online casino has just opened and is making a promotional offer. Each of the first 2 500 online bets of $10 on red at roulette gets back $5 when the bet is lost and $20 when the bet is won. A bet on red is lost with probability $\frac{19}{37}$ and is won with probability $\frac{18}{37}$. Use the central limit theorem to approximate the probability that the casino will lose no more than $6 500 on the promotional offer. (answer: 0.978)

3.6 More on probability densities

First, we discuss in more detail the tricky and subtle concept of probability density function. A random variable X is said to be *continuously distributed* with *probability density function* $f(x)$ if

$$P(X \leq x) = \int_{-\infty}^{x} f(y)\,dy \quad \text{for all real numbers } x,$$

where $f(x)$ is a non-negative function with $\int_{-\infty}^{\infty} f(x)\,dx = 1$ and $f(x)$ is continuous with the possible exception of a finite number of points.[20] The number $f(x)$ is not a probability but measures how densely the probability mass of X is smeared out around a continuity point x:

$$P(x < X \leq x + \Delta x) \approx f(x)\Delta x \quad \text{for } \Delta x \text{ close to } 0.$$

This follows from $P(x < X \leq x + \Delta x) = P(X \leq x + \Delta x) - P(X \leq x)$ and the fact that $f(x)$ is the derivative of $P(X \leq x)$ (if $g(x)$ is the derivative of $G(x)$, then $G(x + \Delta x) - G(x) \approx g(x)\Delta x$ for Δx close to 0). The density function $f(x)$ specifies how the probability mass of the continuously distributed random variable X is smeared out, as were it liquid mass, over the range of the possible values of X. Note

---

[20]In general, you find the probability density function of a continuous random variable X by determining $F(x) = P(X \leq x)$ and differentiating $F(x)$.

that a density function can have values larger than 1, though it must integrate to 1.

In view of the interpretation of $f(x)\Delta x$ for very small Δx, it is reasonable to define the expected value and variance of X as

$$E(X) = \int_{\infty}^{\infty} x f(x)\, dx \quad \text{and} \quad \text{var}(X) = \int_{-\infty}^{\infty} [x - E(X)]^2 f(x)\, dx,$$

assuming that the integrals are well-defined and are finite. Using the formula $P(a < X \le b) = P(X \le b) - P(X \le a)$ and the integral representation of $P(X \le x)$, you have

$$P(a < X \le b) = \int_a^b f(x)\, dx \quad \text{for } a < b.$$

The integral $\int_a^b f(x)\, dx$ is the area under the graph of the density function $f(x)$ between the points a and b. This area goes to 0 if a tends to b. Thus, each individual point has probability mass *zero* for the random variable X and so $P(a < X < b) = P(a < X \le b) = P(a \le X < b) = P(a \le X \le b)$.

3.6.1 The uniform and the beta densities

Let the random variable X be a random point in the finite interval (a, b), where a random point means that the probability of X falling in a sub-interval of (a, b) is proportional to the length of the sub-interval (think of throwing blindly a dart with an infinitely thin point on the interval (a, b)). The proportionality constant must be $\frac{1}{b-a}$ (why?). In particular, for any $a < x < b$, the probability of X falling in the interval (a, x) is $(x - a)/(b - a)$. Thus,

$$P(X \le x) = \frac{x - a}{b - a} \quad \text{for } a < x < b,$$

with $P(X \le x) = 0$ for $x \le a$ and $P(X \le x) = 1$ for $x \ge b$. The derivative of $P(X \le x)$ on (a, b) is

$$f(x) = \frac{1}{b - a} \quad \text{for } a < x < b.$$

Defining $f(x) = 0$ for $x \notin (a, b)$, it follows that $f(x)$ is the probability density function of the random point X. This density function is called the *uniform density function* on (a, b). The probability mass of X is evenly spread out over the interval (a, b). The uniform density function underlies the so-called random numbers in computer simulation. It is a matter of simple algebra to verify that

$$E(X) = \frac{a+b}{2} \quad \text{and} \quad \text{var}(X) = \frac{1}{12}(b-a)^2.$$

The beta density

The uniform density on $(0, 1)$ is a special case of the beta density. The class of beta densities is much used in Bayesian analysis. To introduce the beta density, imagine that you have a biased coin with an unknown probability of coming up heads. Suppose that you model your ignorance about the true value of this probability by a random variable having the uniform density on $(0, 1)$ as probability density. In other words, the prior density $f(p)$ of your belief about the probability of coming up heads is $f(p) = 1$ for $0 < p < 1$. How does your belief about the probability of coming up heads change when you have tossed the coin n times with heads coming up s times and tails $r = n - s$ times? Denote by $f(p \mid s$ heads$)$ the posterior density of your belief about the probability of coming up heads. This density satisfies the Bayes formula

$$f(p \mid s \text{ heads}) = \frac{f(p)\, L(s \text{ heads} \mid p)}{\int_0^1 f(\theta)\, L(s \text{ heads} \mid \theta)\, d\theta},$$

where the likelihood $L(s$ heads $\mid p)$ is the probability of getting s heads and r tails in $r + s$ tosses of the coin when the probability of coming up heads is p. You are asked to take this formula for granted, see also subsection 2.4.2. The essence of the formula is that the posterior density is proportional to the product of the prior density and the likelihood function. In the coin-tossing example the likelihood function is given by the binomial probability

$$L(s \text{ heads} \mid p) = \binom{r+s}{s} p^s (1-p)^r \quad \text{for } 0 < p < 1,$$

and so, with $f(p) = 1$ for $0 < p < 1$, you get the posterior density

$$f(p \mid s \text{ heads}) = \frac{p^s (1-p)^r}{\int_0^1 \theta^s (1-\theta)^r \, d\theta} \quad \text{for } 0 < p < 1.$$

This density is the beta $(s+1, r+1)$ density. In general, the beta (α, β) density with parameters $\alpha > 0$ and $\beta > 0$ has the interval $(0, 1)$ as its range and is defined by

$$\boxed{\frac{1}{B(\alpha, \beta)} x^{\alpha-1} (1-x)^{\beta-1} \quad \text{for } 0 < x < 1,}$$

where $B(\alpha, \beta) = \int_0^1 x^{\alpha-1} (1-x)^{\beta-1} \, dx$. Using induction, it can be shown that $B(\alpha, \beta)$ equals $(\alpha - 1)!(\beta - 1)!/(\alpha + \beta - 1)!$ if α and β are integer-valued. The expected value and the variance of a random variable X with a beta (α, β) density are given by

$$\boxed{E(X) = \frac{\alpha}{\alpha + \beta} \quad \text{and} \quad \text{var}(X) = \frac{\alpha \beta}{(\alpha + \beta)^2 (\alpha + \beta + 1)}.}$$

The uniform density on $(0, 1)$ is the beta (α, β) density with $\alpha = \beta = 1$. A closer look at the above derivation shows that you would have found the beta $(s + \alpha, r + \beta)$ density as posterior density for the probability of coming up heads if you had taken the beta (α, β) density as prior density for this probability. As an illustration, assuming the uniform density as prior density for the probability of coming up heads and tossing the coin 10 times, then the posterior density becomes the beta $(8, 4)$ density when heads appears $s = 7$ times and tails $r = 3$ times. If another 10 tosses are done and these tosses result in 6 heads and 4 tails, the posterior density becomes the beta $(14, 8)$ density. In Bayesian probability, the prior is said to be conjugate for the likelihood when the posterior distribution is in the same family of distributions as the prior belief, but with new parameter values, which have been updated to reflect what has been learned from the data.

3.6.2 The exponential density

The density function of continuous random variable X is said to be an *exponential density function* with parameter $\lambda > 0$ if

$$\boxed{f(x) = \lambda e^{-\lambda x} \quad \text{for } x \geq 0,}$$

and $f(x) = 0$ for $x < 0$. A random variable X with this density can only take on positive values. The exponential probability distribution is often used to model the time until the occurrence of a rare event (e.g., serious earthquake, the decay of a radioactive particle). Figure 10 gives the histogram of a large number of observations of the time until the decay of a radioactive particle. An exponential density function can indeed be very well fitted to this histogram.

The probability distribution function $P(X \le x)$ is (verify!)

$$P(X \le x) = 1 - e^{-\lambda x} \quad \text{for } x \ge 0.$$

A basic formula in integral calculus is $\int_0^\infty x^k e^{-\lambda x}\, dx = \frac{k!}{\lambda^{k+1}}$ for $k = 0, 1, \ldots$ and $\lambda > 0$. Using this formula, you can readily verify that

$$E(X) = \int_0^\infty x f(x)\, dx = \frac{1}{\lambda}, \quad \text{var}(X) = \int_0^\infty \left(x - \frac{1}{\lambda}\right)^2 f(x)\, dx = \frac{1}{\lambda^2}.$$

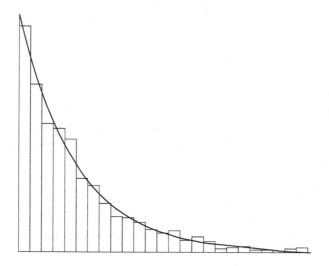

Figure 10: A histogram for decay times.

A characteristic property of the exponential random variable X is its *lack of memory*. That is, for any $s > 0$,

$$P(X > s + x \mid X > s) = P(X > x) \quad \text{for all } x > 0.$$

In other words, imagine that X represents the lifetime of an item, the residual life of the item has the same exponential distribution as the original lifetime, regardless of how long the item has already been in use ('used is as good as new'). The proof is very simple. Since $P(X > v) = e^{-\lambda v}$ for all $v > 0$, you find for any $s > 0$ that

$$P(X > s + x \mid X > s) = \frac{P(X > s + x \text{ and } X > s)}{P(X > s)}$$

$$= \frac{P(X > s + x)}{P(X > s)} = \frac{e^{-\lambda(s+x)}}{e^{-\lambda s}} = e^{-\lambda x} \quad \text{for all } x > 0.$$

The exponential distribution can be shown to be the only continuous distribution having the memoryless property. This property is crucial in the following example.

Example 3.8. You wish to cross a one-way traffic road on which cars drive at a constant speed and pass according to independent inter-arrival times having an exponential distribution with an expected value of $1/\lambda$ seconds. You can only cross the road when no car has come round the corner since c seconds. What is the probability distribution of the number of passing cars before you can cross the road when you arrive at an arbitrary moment?

Solution. Imagine that the time axis is divided in segments of c seconds measured from the moment you arrive at the road. The probability p that the time between the passing of two consecutive cars is more than c seconds is $\int_c^\infty \lambda e^{-\lambda t}\, dt = e^{-\lambda c}$. By the lack of memory of the exponential distribution, $p = e^{-\lambda c}$ gives also the probability that no car comes around the corner during any time segment of c seconds, independently of other time segments. Denoting by the random variable X the number of passing cars before you can cross the road, you now get the shifted geometric distribution

$$P(X = k) = (1 - p)^k\, p \quad \text{for } k = 0, 1, \ldots.$$

Problem 3.37. A satellite has a lifetime that is exponentially distributed with an expected value of 15 years. The satellite is in use for already 12 years. What are the expected value and the standard deviation of the residual lifetime of the satellite? (answer: 15 years) What is the probability that the satellite will survive for another 10 years? (answer: 0.5134)

Problem 3.38. You go by bus to work. It takes 5 minutes to walk from home to the bus stop. To get to work on time, you must take a bus no later than 7:45 a.m. The independent inter-arrival times of the buses are exponentially distributed with a mean of 10 minutes. What is the latest time you must leave home to be on time for work with a probability of at least 0.95? (answer: 7:10 a.m.)

Problem 3.39. The lifetime X of an item is exponentially distributed with expected value $\frac{1}{\lambda}$. Verify that the lifetime has a constant failure rate, that is, $P(t < X \le t + \Delta t \mid X > t) \approx \lambda \Delta t$ for any $t \ge 0$ if Δt is close to zero.

3.7 The Poisson process

The Poisson process links the discrete Poisson distribution and the continuous exponential distribution. Suppose events (e.g. emission of radioactive particles) occur one at a time, where the inter-occurrences times are independent random variables having a same *exponential distribution* with expected value $\frac{1}{\lambda}$. For any $t > 0$, define the random variable $N(t)$ as

$$N(t) = \text{the number of events occurring in } (0, t],$$

where $N(0) = 0$. The random process $\{N(t), t \ge 0\}$ is called a *Poisson process* with rate λ. It can be shown that, for any $s \ge 0$,

$$P(N(s + t) - N(s) = k) = e^{-\lambda t} \frac{(\lambda t)^k}{k!} \quad \text{for } k = 0, 1, \dots \text{ and } t > 0,$$

independently of what happened before time s. Thus, the Poisson process is memoryless, and the number of events occurring in any time interval of length t has a Poisson distribution with expected value λt.

As shown by the Russian mathematician Alexsandr Khinchin (1894–1959), the Poisson process arises only if the following conditions are satisfied:

• Events occur one at a time, that is, two or more events in a very small time interval is practically impossible.
• The numbers of events in non-overlapping time intervals are independent of one another (no after-effects).
• The probability distribution of the number of events occurring during any finite time interval depends only on the length of the interval and not on its position on the time axis.

Example 3.9. In a traditional match between two university soccer teams, goals are scored according to a Poisson process at a rate of $\lambda = \frac{1}{30}$ per minute. The playing time of the match is 90 minutes. What is the probability of having three or more goals during the match? What is the probability that exactly two goals will be scored in the first half of the match and exactly one goal in the second half?

Solution. The number of goals scored during the match has a Poisson distribution with an expected value of $\lambda \times 90 = 3$. Therefore, the probability of having three or more goals during the match is

$$1 - \sum_{k=0}^{2} e^{-3} \frac{3^k}{k!} = 0.5768.$$

By the memoryless property of the Poisson process, the number of goals scored in the first half of the match and the number of goals scored in the second half are independent and have each a Poisson distribution with an expected value of $\lambda \times 45 = 1.5$. Thus, the probability of two goals in the first half of the match and one goal in the second half is given by

$$e^{-1.5} \frac{1.5^2}{2!} \times e^{-1.5} \frac{1.5}{1!} = 0.0840.$$

Example 3.10. A piece of radioactive material emits particles according to a Poisson process with a rate of 0.84 particles per second. A counter detects each emitted particle with probability 0.95, independently of any other particle. In a 10-second period, 12 particles were detected. What is the probability that more than 15 particles were emitted in that period?

Solution. The number of particles that will be emitted during a 10-second period has a Poisson distribution with expected value $10 \times 0.84 = 8.4$. Using the Poissonization result in Section 3.3, the number of emitted particles that will be missed by the counter in the 10-second period has a Poisson distribution with expected value $0.05 \times 8.4 = 0.420$. The sought probability is the probability of having more than three emissions of undetected particles in the 10-second period. This probability is

$$1 - \sum_{j=0}^{3} e^{-0.420} \frac{0.420^j}{j!} = 0.00093.$$

Merging and splitting of Poisson processes

In general, random splitting of a Poisson process leads to separate, independent Poisson processes. Also, merging independent Poisson processes leads to a Poisson process, see also Example 2.18. In mathematical terms, if $\{N_i(t), t \geq 0\}$ is a Poisson process in which type-i events occur at a rate of λ_i per unit time for $i = 1, 2$, then the merged process $\{N_1(t) + N_2(t), t \geq 0\}$ is a Poisson process in which events occur at a rate of $\lambda_1 + \lambda_2$ per unit time and each event is a type i event with probability $\frac{\lambda_i}{\lambda_1 + \lambda_2}$, independently of any other event. On the other hand, if $\{N(t), t \geq 0\}$ is a Poisson process with rate λ and any event occurring in this process gets label j with probability p_j for $j = 1, \ldots, r$, independently of any other event, then the split-off process $\{N_j(t), t \geq 0\}$ of events with label j is a Poisson process with rate λp_j, and these r Poisson processes are independent.

The Poisson process and the uniform distribution

In a Poisson process, events occur completely randomly in time. This fact is reflected in the following relationship between a Poisson process and the uniform distribution: given that exactly n events have occurred in a given time interval $(0, t)$, the joint distribution of the epochs of the n events is the same as the joint distribution of the order statistics of n independent random variables that are uniformly distributed on $(0, t)$. Randomly chosen points in an interval are not evenly distributed over the interval (otherwise, they would not be

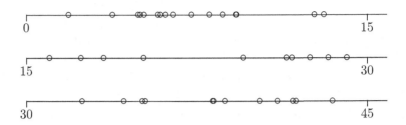

Figure 11: Simulated events in a Poisson process.

random), but tend to cluster. Therefore, a Poisson process also has the property that the epochs at which events occur tend to cluster. For example, this property may be used to explain the surprisingly large number of shark attacks in Florida in the summer of 1991. Unpredictable events such as shark attacks can be modeled by a Poisson process with its bursty behavior. This means that there are periods with a much higher than average number of attacks, as well as periods with no attacks at all. The clustering phenomenon is illustrated in Figure 11, which gives the simulated epochs of events in the time interval (0,45) for a Poisson process with rate $\lambda = 1$. How to simulate a Poisson process will be discussed in subsection 5.2.6.

Problem 3.40. Major cracks on a highway occur according to a Poisson process at a rate of one per 10 miles. What is the probability of two or more major cracks on a specific 15-mile stretch of the highway? (answer: 0.4422)

Problem 3.41. In any 20-minute interval, there is a 10% probability of seeing at least one shooting star. What is the probability of seeing at least one shooting star in the period of an hour? (answer: 0.271)

Problem 3.42. In a video game with a time slot of fixed length T, signals occur according to a Poisson process with rate λ, where $T > \frac{1}{\lambda}$. In the time slot you can push a button only once. You win if at least one signal occurs in the time slot, and you push the button at the occurrence of the last signal. Your strategy is to let pass a fixed time s and push the button upon the first occurrence of a signal (if any) after time s. What value of s maximizes the probability of winning the game? (answer: $s = T - \frac{1}{\lambda}$) What is the maximum probability of winning the game? (answer: $\frac{1}{e}$)

Problem 3.43. The amount of time needed to wash a car at a washing station is exponentially distributed with an expected value of 15 minutes. The washing station can handle only one car at a time. You arrive at the washing station while it is occupied, and one other car is waiting for a washing. The owner of this car informs you that the car in the washing station is already there for 10 minutes. What is the probability that you have to wait more than 20 minutes before your car can be washed? (answer: 0.6151)

Problem 3.44. On Wednesday afternoon between 1 p.m. and 4:30 p.m., buses with tourists arrive in Gotham city to visit the castle in this picturesque town. The times between successive arrivals of buses are independent random variables each having an exponential distribution with an expected value of 45 minutes. Each bus stays exactly 30 minutes on the parking lot of the castle. The parking lot has ample space. What is the probability mass function of the number of buses on the parking lot at 4 p.m.? (answer: Poisson distribution with expected value $\frac{2}{3}$).

3.8 The Q-Q plot and the chi-square test

In this section the chi-square test along with the visual assessment tool of the Q-Q plot will be discussed. The Q-Q plot is used to get a first idea of the shape of the distribution underlying a set of independent data and the chi-square test is used to determine whether there is a statistically significant difference between the observed frequencies and the expected frequencies of the data.

Q-Q plot

In general, the Q-Q plot can be used when you have data independently drawn from the same underlying probability distribution and you wish to get some idea of the unknown underlying distribution. Before you estimate the parameters of the population distribution, you should decide what general family of distributions is appropriate for the data. Does the underlying distribution belong to the family of, for example, the normal, uniform or exponential distributions? The procedure for the Q-Q plot is as follows. Suppose the data $x_1, \ldots, x_n$ are independently drawn from a continuous probability distribution. The first step is to order the data according to

$$x_{(1)} < x_{(2)} < \cdots < x_{(n)}.$$

The $x_{(i)}$ are called order statistics. The proportion of the data that are smaller than or equal to $x_{(i)}$ is $\frac{i}{n}$. That is, $x_{(i)}$ can be seen as the $\frac{i}{n}$th percentile of the empirical distribution of the data. For technical reasons, it is more convenient to consider $x_{(i)}$ as an $\frac{i-0.5}{n}$ percentile. The next step is to hypothesize some distributional form for the data, and then compare percentiles. The pth percentile η_p of a continuous random variable X with a strictly increasing probability distribution function $F(x) = P(X \leq x)$ is defined as the unique solution of $F(x) = p$, and so

$$\eta_p = F^{-1}(p) \quad \text{for } 0 < p < 1.$$

The empirical $\frac{i-0.5}{n}$ percentiles $x_{(i)}$ are now compared with the hypothesized $\frac{i-0.5}{n}$ percentiles $F^{-1}\left(\frac{i-0.5}{n}\right)$ for $i = 1, \ldots, n$. In the Q-Q plot the points

$$\left(x_{(i)}, F^{-1}\left(\frac{i - 0.5}{n}\right)\right)$$

are plotted for $i = 1, \ldots, n$ and you look for linearity. If the points lie closely on a straight line, you have strong (but not conclusive) evidence that the data are independently drawn from the hypothesized distribution. It is noted that for some hypothesized distributional forms you don't need to know the parameters of the distribution for the Q-Q plot. The standard normal distribution suffices, the uniform distribution on $(0, 1)$, and the exponential distribution with scale parameter 1. By doing so, you only apply a linear transformation on the parameters of these three distributions. You should realize that the Q-Q plot is only an exploratory tool which gives you a first idea about the distribution. After you have chosen the distributional form for the data, you can estimate the parameters of the distribution from the data and apply the chi-square test to find out how well the fitted distribution agrees with the data.

Chi-square test

The chi-square test (χ^2 test) is one of the most useful statistical tests. It is used to test whether data were generated from a particular probability distribution. The test can also be used to assess whether data has been manipulated to bring observed frequencies closer to expected frequencies.

Suppose you want to find out whether the probability mass function $p_1, \ldots, p_r$ fits a random sample of observations obtained for a repeatable chance experiment with a finite number of possible outcomes $O_1, \ldots, O_r$. To introduce the chi-square test, denote by the random variable N_j the number of times that outcome O_j will appear in n physically independent repetitions of the chance experiment in which outcome O_j occurs with probability p_j. The random variable N_j has a binomial distribution with parameters n and p_j, and so the expected value of N_j is np_j. By the principle of least squares, it is reasonable to consider a test statistic of the form $\sum_{j=1}^{r} w_j(N_j - np_j)^2$ for appropriately chosen weights w_j. It turns out that the choice $w_j = \frac{1}{np_j}$ yields a statistic with a tractable distribution. Thus, the so-called chi-square statistic is defined by

$$D = \sum_{j=1}^{r} \frac{(N_j - np_j)^2}{np_j}.$$

The probability distribution of the statistic D is difficult to compute. However, the discrete probability distribution of D can be very accurately approximated by a tractable continuous distribution when np_j is sufficiently large for all j, say $np_j \geq 5$ for all j (in order to achieve this, it might be necessary to pool some data groups). Then,

$$P(D \leq x) \approx P(\chi^2_{r-1} \leq x) \quad \text{for } x \geq 0,$$

where the continuous random variable χ^2_{r-1} is distributed as the sum of the squares of $r - 1$ independent $N(0, 1)$ distributed variables. The probability distribution of χ^2_{r-1} is called the chi-square distribution with $r - 1$ degrees of freedom. Its expected value is

$$E(\chi^2_{r-1}) = r - 1.$$

It should be pointed out that the above approximation for the chi-square statistic D assumes that the probabilities p_j are not estimated from the data but are known beforehand; if you have to estimate one or more parameters to get the probabilities p_j, you must lower the number of degrees of freedom of the chi-square distribution by one for every parameter estimated from the data.

How do you apply the chi-square test in practice? Using the data that you have obtained for the chance experiment in question, you calculate the numerical value d of the test statistic D for these data. The (subjective) judgment whether the probability mass function $p_1, \ldots, p_r$ fits the data depends on the value of $P(D \leq d)$.

Example 3.11. Somebody claims to have rolled a fair die 1 200 times and to have found that the outcomes 1, 2, 3, 4, 5, and 6 occurred 196, 202, 199, 198, 202, and 203 times. Do you believe these results?

Solution. The reported frequencies are very close to the expected frequencies. Since the expected value and the standard deviation of the number of rolls with outcome j are $1\,200 \times \frac{1}{6} = 200$ and $\sqrt{1\,200 \times (1/6) \times (5/6)} = 12.91$ for all j, you should be suspicious about the reported results. You can substantiate this with the chi-square test. The chi-square statistic D takes on the value

$$\frac{1}{200}\big[(196 - 200)^2 + (202 - 200)^2 + (199 - 200)^2 + (198 - 200)^2$$
$$+ (202 - 200)^2 + (203 - 200)^2\big] = 0.19.$$

The value 0.19 lies far below the expected value 5 of the chi-square distribution with $6 - 1 = 5$ degrees of freedom. The probability $P(D \leq 0.19)$ is approximated by

$$P(\chi_5^2 \leq 0.19) = 0.00078.$$

The simulated value of $P(D \leq 0.19)$ is 0.00083 (four million simulation runs). The very small probability for the test statistic indicates that the data are indeed most likely fabricated.

Example 3.12. A total of 64 matches were played during the World Cup soccer 2010 in South Africa. There were 7 matches with zero goals, 17 matches with 1 goal, 13 matches with two goals, 14 matches with three goals, 7 matches with four goals, 5 matches with five goals, and 1 match with seven goals. Does a Poisson distribution fit these data?

Solution. In this example you must first estimate the unknown parameter λ of the hypothesized Poisson distribution. The parameter λ is estimated by

$$\frac{1}{64}\left(17 \times 1 + 13 \times 2 + 14 \times 3 + 7 \times 4 + 5 \times 5 + 0 \times 6 + 1 \times 7\right)$$

and so $\lambda = \frac{145}{64}$. In order to satisfy the requirement that each data group should have an expected size of at least 5, the matches with 5 or more goals are aggregated, and so six data groups are considered. If a Poisson distribution with expected value $\lambda = \frac{145}{64}$ applies, then the expected number of matches with exactly j goals is $64 \times e^{-\lambda} \lambda^j / j!$ for $j = 0, 1, \ldots, 4$ and the expected number of matches with 5 or more goals is $64 \times (1 - \sum_{j=0}^{4} e^{-\lambda} \lambda^j / j!)$. These expected numbers have the values 6.641, 15.046, 17.044, 12.872, 7.291, and 5.106. Thus, the value of the chi-square test statistic D is given by

$$\frac{(7 - 6.641)^2}{6.641} + \frac{(17 - 15.046)^2}{15.046} + \frac{(13 - 17.044)^2}{17.044} + \frac{(14 - 12.872)^2}{12.872}$$
$$+ \frac{(7 - 7.291)^2}{7.291} + \frac{(6 - 5.106)^2}{5.106} = 1.500.$$

Since the parameter λ was estimated from the data, the test statistic D has approximately a chi-square distribution with $6 - 1 - 1 = 4$ degrees of freedom. By

$$P(\chi_4^2 \geq 1.500) = 0.827,$$

the Poisson distribution gives an excellent fit to the data.

Problem 3.45. In a classical study on the distribution of 196 soldiers kicked to death by horses among 14 Prussian cavalry corps over the 20 years from 1875 to 1894, the data are as follows. In 144 corps-years no deaths occurred, 91 corps-years had one death, 32 corps-years had two deaths, 11 corps-years had three deaths, and

2 corps-years had four deaths. Does a Poisson distribution fit the data? (answer: yes, $P(\chi_2^2 \geq 1.952) = 0.377$)[21]

Problem 3.46. In a famous physics experiment done by Rutherford, Chadwick, and Ellis in 1920, the number of α-particles emitted by a piece of radioactive material were counted during 2608 time intervals of each 7.5 seconds. Denoting by O_j the number of intervals with exactly j particles, the observed data are $O_0 = 57$, $O_1 = 203$, $O_2 = 383$, $O_3 = 525$, $O_4 = 532$, $O_5 = 408$, $O_6 = 273$, $O_7 = 139$, $O_8 = 45$, $O_9 = 27$, $O_{10} = 10$, $O_{11} = 4$, $O_{12} = 0$, $O_{13} = 1$, and $O_{14} = 1$. Do the observed frequencies conform to Poisson frequencies? (answer: yes, $P(\chi_{10}^2 \geq 12.961) = 0.226$)

3.9 The bivariate normal density

Multivariate normal distributions are the most important joint distributions of two or more random variables. This section deals with the bivariate normal distribution. An addendum briefly discusses the general case of two jointly distributed continuous random variables.

A random vector (X, Y) is said to have a *standard bivariate normal distribution* with parameter ρ if

$$P(X \leq x \text{ and } Y \leq y) = \int_{-\infty}^{x} \int_{\infty}^{y} \varphi(v, w)dw\, dv \text{ for } -\infty < x, y < \infty,$$

where $\varphi(x, y)$ is the *standard bivariate normal density function*

$$\varphi(x, y) = \frac{1}{2\pi\sqrt{1-\rho^2}}\, e^{-\frac{1}{2}\left(x^2 - 2\rho xy + y^2\right)/(1-\rho^2)} \qquad \text{for all } x \text{ and } y.$$

The parameter ρ is a constant with $-1 < \rho < 1$. Figure 12 shows the characteristic shape of the bivariate normal density. The density function $\varphi(x, y)$ allows for the interpretation: the probability that the pair (X, Y) will take on a value in a small rectangle around the point (x, y) with side lengths Δx and Δy is approximately $\varphi(x, y)\Delta x \Delta y$.

---

[21]This study was done by the Russian statistician Ladislaus von Bortkiewicz (1868–1931), who first discerned and explained the importance of the Poisson distribution in his book *Das Gesetz der Kleinen Zahlen*. The French mathematician Siméon-Denis Poisson (1781–1840) himself did not recognize the huge practical importance of the distribution that would later be named after him. By the way, this distribution was first found by Abraham de Moivre (1667–1754) in 1711.

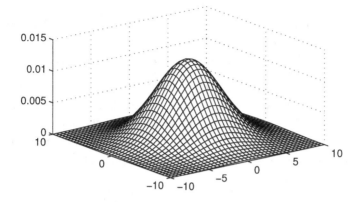

Figure 12: A bivariate normal density.

For a pair (X, Y) having a standard bivariate normal distribution with parameter ρ, it is a matter of integral calculus to show that

- both X and Y are $N(0, 1)$ distributed;[22]

- the correlation coefficient of X and Y is ρ.

More generally, the vector (X, Y) is said to have a bivariate normal density with parameters $(\mu_X, \mu_Y, \sigma_X^2, \sigma_Y^2, \rho)$ if the standardized vector $\left(\frac{X-\mu_X}{\sigma_X}, \frac{Y-\mu_Y}{\sigma_Y}\right)$ has the standard bivariate normal density with parameter ρ. Then, by taking the partial derivatives of $P(X \leq x$ and $Y \leq y) = \int_0^{(x-\mu_X)/\sigma_X} \int_0^{(y-\mu_Y)/\sigma_Y} \varphi(v, w) dw \, dv$ with respect to x and y, it follows that the joint density $f_{X,Y}(x, y)$ of the vector (X, Y) is equal to

$$\frac{1}{2\pi\sigma_X\sigma_Y\sqrt{1-\rho^2}} e^{-\frac{1}{2}\left[\left(\frac{x-\mu_X}{\sigma_X}\right)^2 - 2\rho\left(\frac{x-\mu_X}{\sigma_X}\right)\left(\frac{y-\mu_Y}{\sigma_Y}\right) + \left(\frac{y-\mu_Y}{\sigma_Y}\right)^2\right]/(1-\rho^2)}.$$

---

[22]The proof is based on $P(X \leq x) = \int_{-\infty}^x \left(\int_{-\infty}^\infty \varphi(v, w) dw\right) dv$. Differentiation gives that the density of X is $\int_{-\infty}^\infty \varphi(x, w) \, dw$. You can decompose $\varphi(x, w)$ as $(1/\sqrt{2\pi}) e^{-\frac{1}{2}x^2} \times \varphi(w \mid x)$ with $\varphi(w \mid x) = \left(1/(\sqrt{1-\rho^2}\sqrt{2\pi})\right) e^{-\frac{1}{2}(w-\rho x)^2/(1-\rho^2)}$. Noting that $\varphi(w \mid x)$ is an $N(\rho x, 1-\rho^2)$ density for fixed x and thus integrates to 1, you get that $(1/\sqrt{2\pi})e^{-\frac{1}{2}x^2}$ is the marginal density of X. Since $E(XY) = \int_{-\infty}^\infty \int_{-\infty}^\infty xy \, \varphi(x, y) dy \, dx$, the decomposition of $\varphi(x, y)$ leads to $E(XY) = \rho$, which gives that the correlation coefficient $\rho(X, Y) = \rho$.

The random variables X and Y are $N(\mu_X, \sigma_X^2)$ and $N(\mu_Y, \sigma_Y^2)$ distributed, and their correlation coefficient is ρ. A characteristic property of the bivariate normal distribution is that X and Y are independent if $\rho = 0$ (the converse is always true).

In general, a linear combination of normally distributed random variables is not normally distributed. However, it can be shown that the random vector (X, Y) has a bivariate normal distribution if and only if any linear combination of X and Y is normally distributed. This alternative definition is useful when generalizing to the multivariate normal distribution.

An important concept is the *conditional density function* of Y given that $X = x$. This density function is denoted by $f_Y(y \mid x)$ for fixed x and can be defined through the product rule

$$\boxed{f_{X,Y}(x, y) = f_X(x) f_Y(y \mid x),}$$

where the $N(\mu_X, \sigma_X^2)$ density $f_X(x)$ is the probability density of X. Similarly, the conditional density function $f_X(x \mid y)$ of X is defined. The definition of the conditional densities parallels the formula

$$P(V = v \text{ and } W = w) = P(V = v)P(W = w \mid V = v)$$

for discrete random variables V and W. The conditional density function $f_Y(y \mid x)$ can be shown to be a normal density function whose expected value and variance are given by

$$\boxed{E(Y \mid X = x) = \mu_Y + \rho \frac{\sigma_Y}{\sigma_X}(x - \mu_X) \quad \text{for all } x}$$

and

$$\boxed{\sigma^2(Y \mid X = x) = (1 - \rho^2)\sigma_Y^2 \quad \text{for all } x.}$$

Thus, a random vector (X, Y) with a bivariate normal distribution has the nice property that the optimal predictor $E(Y \mid X = x)$ is given by the *linear least squares regression line*, see also Problem 2.62. The predictor $E(Y \mid X = x)$ minimizes $E[(Y - g(X))^2]$ over all functions $g(x)$. Note that the residual variance $(1 - \rho^2)\sigma_Y^2$ of Y does not depend on x.

Regression to the mean

Linear regression attempts to model the relationship between two variables by fitting a linear equation to observed data. One variable is considered to be an explanatory variable, and the other is considered to be a dependent variable. In a classical study on the heights of fathers and their adult sons at the same age, Sir Francis Galton (1822–1911) measured the heights of 1078 fathers and sons. Galton observed that tall fathers tend to have somewhat shorter sons and short fathers somewhat taller sons. The phenomenon of *regression to the mean* can be explained with the help of the regression line. Think of X as the height of a 25-year-old father and think of Y as the height his newborn son will have at the age of 25 years. If (X, Y) has a bivariate normal distribution such that $\mu_X = \mu_Y = \mu$ and $\sigma_X = \sigma_Y = \sigma$, the best prediction of Y given that $X = x$ is $E(Y \mid X = x) = \mu + \rho(x - \mu)$. Thus, if the observed height x of the father scores above the mean μ and the correlation ρ between X and Y is positive, you get $0 < E(Y \mid X = x) - \mu < x - \mu$. That is, the best linear prediction is that the height of the son will score closer to the mean than the height of the father. Very tall fathers tend to have somewhat shorter sons and very short fathers somewhat taller ones! Regression to the mean shows up in a wide variety of places: it helps explain why great movies have often disappointing sequels, and disastrous presidents have often better successors.

Multiple linear regression

In multiple regression, you want to predict the value of a response variable Y based on the values of explanatory variables $X_1, \ldots, X_s$. Let's assume that $(Y, X_1, \ldots, X_s)$ has a multivariate normal distribution, that is, each linear combination of these $s + 1$ variables is normally distributed. Then it can be shown that

$$E(Y \mid X_j = x_j \text{ for } j = 1, \ldots, s) = \theta_0 + \theta_1 x_1 + \cdots + \theta_s x_s.$$

The parameters $\theta_0, \theta_1, \ldots, \theta_s$ can be estimated by minimizing

$$J(\boldsymbol{\theta}) = \sum_{i=1}^{n} \left[y^{(i)} - (\theta_0 + \theta_1 x_1^{(i)} + \cdots + \theta_s x_s^{(i)}) \right]^2$$

when n independent observations $\left(y^{(i)}, x_1^{(i)}, \ldots, x_s^{(i)}\right)$ are available for large n. As done for logistic regression in subsection 2.10.3, gradient descent can be applied to minimize $J(\boldsymbol{\theta})$.

Problem 3.47. A statistics class has two exams, and the scores of the students on the exams 1 and 2 follow a bivariate normal distribution with parameters $\mu_1 = 75$, $\mu_2 = 65$, $\sigma_1 = 12$, $\sigma_2 = 15$, and $\rho = 0.7$. Take a randomly chosen student. What is the probability that the score on exam 1 is 80 or more? (answer: 0.3385) What is the probability that total score over the two exams will exceed 150? (answer: 0.3441) What is the probability that the student will do better on exam 2 than on exam 1? (answer: 0.1776) What is the probability that the score on exam 2 will be over 80 given a score of 80 on exam 1? (answer: 0.1606)

Problem 3.48. Suppose that the joint distribution of the heights of fathers and their adult sons at the same age can be modeled by a bivariate normal density with $\mu_X = 67.7$, $\mu_Y = 68.7$, $\sigma_X = 2.7$, $\sigma_Y = 2.7$, and $\rho = 0.5$. What is the prediction for the height of the son if the father has a height of 73.1 inches? (answer: 71.4 inches)

3.9.1 Additional material for joint random variables

In the foregoing discussion, key concepts for the joint distribution of two random variables were introduced between the lines. This subsection briefly discusses these concepts again in a more general context.

In general, two continuous random variables X and Y that are defined on a same sample space are said to have a joint probability density function $f_{X,Y}(x, y)$ if

$$P(X \leq a \text{ and } Y \leq b) = \int_{x=-\infty}^{a} \int_{y=-\infty}^{b} f_{X,Y}(x, y)\, dy\, dx$$

for all $-\infty < a, b < \infty$, where $f_{X,Y}(x, y)$ is a *non-negative* function satisfying $\int_{-\infty}^{\infty} \int_{-\infty}^{\infty} f_{X,Y}(x, y)\, dy\, dx = 1$. In specific applications, you usually find the joint density function by determining first the joint cumulative probability distribution function and taking next the second-order partial derivative. Since $P\big((X, Y) \in C\big) = \iint_C f_{X,Y}(x, y)\, dy\, dx$ for any neat region C in the plane, the probability that (X, Y) falls in the region C is the volume under the joint density surface over the region C.

An important joint density is the uniform density. Suppose that a point (X, Y) is picked at random inside a bounded region R in the plane. The joint density of (X, Y) is the *uniform density*

$$f_{X,Y}(x, y) = \frac{1}{\text{area of region } R} \quad \text{for } (x, y) \in R,$$

and $f_{X,Y}(x, y) = 0$ for $(x, y) \notin R$. In particular, a random point inside a circle with radius r has the density $f_{X,Y}(x, y) = \frac{1}{\pi r^2}$ on the circle.

If the random vector (X, Y) has a joint probability density $f_{X,Y}(x, y)$, then each of the random variables X and Y has a probability density itself. These densities are called *marginal densities*. Differentiating

$$P(X \le x) = \lim_{y \to \infty} P(X \le x, Y \le y) = \int_{-\infty}^{x} \left[\int_{-\infty}^{\infty} f_{X,Y}(v, w) \, dw \right] dv$$

gives the marginal density of the random variable X:

$$f_X(x) = \int_{-\infty}^{\infty} f_{X,Y}(x, y) \, dy \quad \text{for } -\infty < x < \infty.$$

As an illustration, let the random variable L be the length of the line segment between the center of a circle with radius r and a randomly picked point inside the circle, and let Θ be the angle between this line segment and the horizontal axis. What are the marginal densities of L and Θ? Since the area of a circle sector with radius x and central angle θ is $\frac{\theta}{2\pi} \times \pi x^2 = \frac{1}{2}\theta x^2$, you get

$$P(L \le x \text{ and } \Theta \le \theta) = \frac{\theta x^2}{2\pi r^2} \quad \text{for } 0 \le x \le r, 0 \le \theta \le 2\pi.$$

Taking the second-order partial derivative, you get

$$f_{L,\Theta}(x, \theta) = \frac{x}{\pi r^2} \quad \text{for } 0 < x < r \text{ and } 0 < \theta < 2\pi.$$

Then, by the formulas $f_L(x) = \int_0^{2\pi} \frac{x}{\pi r^2} \, d\theta$ and $f_\Theta(\theta) = \int_0^r \frac{x}{\pi r^2} \, dx$,

$$f_L(x) = \frac{2x}{r^2} \text{ for } 0 < x < r \quad \text{and} \quad f_\Theta(\theta) = \frac{1}{2\pi} \text{ for } 0 < \theta < 2\pi.$$

It is interesting to note that $f_{L,\Theta}(x,\theta) = f_L(x) \times f_\Theta(\theta)$ for all (x,θ). This means that L and Θ are independent of each other.

Another important concept is that of *conditional density*. The conditional density function of X given that $Y = y$ with $f_Y(y) > 0$ is denoted by $f_X(x \mid y)$ and is defined through the formula

$$\boxed{f_{X,Y}(x,y) = f_Y(y)f_X(x \mid y).}$$

Thus, for any fixed y with $f_Y(y) > 0$, $f_X(x \mid y)$ is defined by

$$\boxed{f_X(x \mid y) = \frac{f_{X,Y}(x,y)}{f_Y(y)} \quad \text{for} \ -\infty < x < \infty.}$$

The conditional expectation of X given that $Y = y$ is defined by

$$\boxed{E(X \mid Y = y) = \int_{-\infty}^{\infty} x f_X(x \mid y)\,dx.}$$

As an illustration, let (X,Y) be a randomly chosen point inside the unit circle with the origin as center. Then, the joint density of (X,Y) is the uniform density $f_{X,Y}(x,y) = \frac{1}{\pi}$ on the circle. By integrating $f_{X,Y}(x,y) = \frac{1}{\pi}$ over x from $-\sqrt{1-y^2}$ to $\sqrt{1-y^2}$, you get the marginal density

$$f_Y(y) = \frac{2}{\pi}\sqrt{1-y^2} \quad \text{for} \ -1 < y < 1.$$

Next, you find for $f_X(x \mid y)$ the uniform density

$$f_X(x \mid y) = \frac{1}{2\sqrt{1-y^2}} \quad \text{for} \ -\sqrt{1-y^2} < x < \sqrt{1-y^2}.$$

You are asked to verify that $E(X \mid Y = y) = 0$ for all $-1 < y < 1$ and $\rho(X,Y) = 0$. But X and Y are not independent!

The two-dimensional version of the substitution rule is:

$$\boxed{E\big[g(X,Y)\big] = \int_{-\infty}^{\infty}\int_{-\infty}^{\infty} g(x,y)f_{X,Y}(x,y)\,dy\,dx}$$

for any function $g(x, y)$ for which the double integral is well-defined. The integral always exists when the function $g(x, y)$ is non-negative. As an illustration, let's calculate the expected value of the distance between a randomly chosen point (X, Y) inside the unit circle and the center of the circle. The expected distance can be written as $\iint_C \sqrt{x^2 + y^2} \frac{1}{\pi} \, dy \, dx$, where $C = \{(x, y) : x^2 + y^2 \leq 1\}$. Using polar coordinates $x = r\cos(\theta)$ and $y = r\sin(\theta)$, this double integral can be computed as

$$\frac{1}{\pi} \int_0^{2\pi} \int_0^1 r \, r dr d\theta = 2 \int_0^1 r^2 \, dr,$$

and so the expected distance between a random point (X, Y) inside the unit circle and the center of the circle is $\frac{2}{3}$.

The basic formula $f_{X,Y}(x, y) = f_Y(y) f_X(x \mid y)$ is very important. It may be helpful in simulating a random observation from the joint density $f_{X,Y}(x, y)$ of the random vector (X, Y): first, an observation y is simulated from the marginal density $f_Y(y)$, and then an observation x from the conditional density $f_X(x \mid y)$.

Also, the formula $f_{X,Y}(x, y) = f_Y(y) f_X(x \mid y)$ is crucial in the proof of the continuous version of the *law of conditional expectation*:

$$E(X) = \int_{-\infty}^{\infty} E(X \mid Y = y) f_Y(y) \, dy.$$

A sketch of the proof is as follows. Using the definitions $f_X(x) = \int_{-\infty}^{\infty} f_{X,Y}(x, y) \, dy$ and $f_{X,Y}(x, y) = f_Y(y) f_X(x \mid y)$, you get

$$E(X) = \int_{-\infty}^{\infty} x f_X(x) \, dx = \int_{-\infty}^{\infty} x \, dx \left[\int_{-\infty}^{\infty} f_Y(y) f_X(x \mid y) \, dy \right]$$

$$= \int_{-\infty}^{\infty} f_Y(y) \, dy \int_{-\infty}^{\infty} x f_X(x \mid y) \, dx = \int_{-\infty}^{\infty} E(X \mid Y = y) f_Y(y) \, dy.$$

Here the assumption is made that the integrals exist and that the interchange of the order of integration is allowed. This assumption is satisfied when X and Y are non-negative random variables. The *law of conditional probability* can be obtained in the same way as the law of conditional expectation:

$$P(X \leq x) = \int_{-\infty}^{\infty} P(X \leq x \mid Y = y) f_Y(y) \, dy \quad \text{for any } x.$$

The laws of conditional expectation and conditional probability are very useful in solving probability problems, where the choice of the conditioning variable Y is usually obvious from the context of the problem.

Example 3.13. When you go home from work, you will arrive at the bus stop at a random time between 5:45 p.m. and 6:00 p.m. Bus numbers 1 and 3 bring you home. You take the first bus that arrives. Bus number 1 arrives exactly every 15 minutes starting from the hour, and the interarrival times of buses number 3 are independent random variables having an exponential distribution with a mean of 15 minutes. What is your expected waiting time until the first arrival of a bus? What is the probability that bus number 1 will arrive first to take you home?

Solution. Let the random variable X be your waiting time (in minutes) until the first arrival of a bus. To calculate $E(X)$, condition on the random variable Y denoting the number of minutes between your arrival time at the bus stop and 6:00 p.m. The random variable Y is uniformly distributed on $(0, 15)$ and has density $f_Y(y) = \frac{1}{15}$ for $0 < y < 15$. Using the lack of memory of the exponential distribution, the conditional expected waiting time $E(X \mid Y = y)$ is the same as the expected value of the random variable $\min(y, V)$, where the inter-arrival time V has the exponential density $\frac{1}{15}e^{-v/15}$. Then, using the substitution rule,

$$E(X \mid Y = y) = E[\min(y, V)] = \int_0^\infty \min(y, v)\frac{1}{15}e^{-v/15}\,dv$$

$$= \int_0^y v\,\frac{1}{15}e^{-v/15}\,dv + y\int_y^\infty \frac{1}{15}e^{-v/15}\,dv$$

$$= 15\big[1 - e^{-y/15} - \frac{y}{15}e^{-y/15}\big] + y\,e^{-y/15} = 15(1 - e^{-y/15}).$$

The law of conditional expectation now gives

$$E(X) = \int_0^{15} E(X \mid Y = y)\frac{1}{15}\,dy = \int_0^{15}\left(1 - e^{-y/15}\right)dy = \frac{15}{e}.$$

The probability that bus number 1 will arrive first to take you home is $\int_0^{15} P(V > y)\frac{1}{15}\,dy = 1 - \frac{1}{e}$, by the law of conditional probability.

Chapter 4

Real-World Applications of Probability

This chapter discusses real-world applications of probability. The Poisson distribution plays a key role in several applications involving rare events. This distribution is a particularly suitable distribution for modeling rare events. It has also the practically useful property that its standard deviation is the square root of its mean. Applications include detecting fraud in a Canadian lottery, real-life cases of the birthday problem, collecting coupons at World Cup soccer tournaments, and Benford's law for detecting potential fraud in financial records.

4.1 Fraud in a Canadian lottery

In the Canadian province of Ontario, a strong suspicion arose at a certain point that winning lottery tickets were repeatedly stolen by lottery employees from people who had their lottery ticket checked at a point of sale. These people, mostly the elderly, were then told that their ticket had no prize and that it could go into the trash. The winning ticket was subsequently surrendered by the lottery ticket seller, who pocketed the cash prize. The ball started to roll when an older participant — who always entered the same numbers on his ticket — found out that in 2001 a prize of $250 000 was taken from him by a sales point employee.

How do you prove a widespread fraud in the lottery system? Statistical analysis by Jeffrey Rosenthal, a well-known Canadian professor

131

of probability, showed the fraud. The Poisson distribution played a major role in the analysis. How was the analysis done? Rosenthal worked together with a major Canadian TV channel. Investigation by the TV channel, making an appeal to the freedom of information act, revealed that in the period 1999–2006 there were 5 713 big prizes ($50 000 or more) of which 200 prizes were won by lottery ticket sellers. Can this be explained as a fluke of chance? To answer this question, you need to know how many people are working at points of sale of the lottery. There were 10 300 sales outlets in Ontario, and research by the TV channel led to an estimated average of 3.5 employees per point of sale, or about 36 thousand employees in total. The lottery organization fought this number and came up with 60 thousand lottery ticket sellers. You also need to know how much the average lottery ticket seller spends on the purchase of lottery tickets compared to the average adult inhabitant of Ontario. The estimate by the TV channel was that the average expenditure on lottery tickets taken over all lottery ticket sellers was about 1.5 times as large as the average expenditure on lottery tickets taken over all 8.9 million adult residents of Ontario.

Let's now calculate the probability that the lottery sellers will win 200 or more of the 5 713 big prizes when there are 60 thousand lottery ticket sellers with an expenditure factor of 1.5. In that case, the expected number of winners of big prizes among the lottery ticket sellers can be estimated as

$$5\,713 \times \frac{60\,000 \times 1.5}{8\,900\,000} = 57.$$

In view of the physical background of the Poisson distribution — the probability distribution of the total number of successes in a very large number of independent trial each having a very small probability of success — it is plausible to use the Poisson distribution to model the number of winners among the lottery ticket sellers. The Poisson distribution has the nice feature that its standard deviation is the square root of its expected value. Moreover, nearly all the probability mass of the Poisson distribution lies within three standard deviations from the expected value when the expected value is not too small, see Section 3.3. Two hundred winners among the lottery ticket sellers lies

$$\frac{200 - 57}{\sqrt{57}} \approx 19$$

standard deviations above the expected value. The probability corresponding to a z-score of 19 is inconceivably small (on the order of 10^{-49}) and makes clear that there is large-scale fraud in the lottery.

The lottery organization objected to the calculations and came with new figures. Does the conclusion of large-scale fraud change for the rosy-tinted figures of 101 000 lottery ticket sellers with an expenditure factor of 1.9? Then you get the estimate

$$5\,713 \times \frac{101\,000 \times 1.9}{8\,900\,000} \approx 123$$

for the expected number of winners under the lottery ticket sellers. Two hundred winners is still

$$\frac{200 - 123}{\sqrt{123}} \approx 7$$

standard deviations above the expected value. A z-score of 7 has also a negligibly small probability (on the order of 10^{-7}) and cannot be explained as a chance fluctuation. It could not be otherwise that there was large-scale lottery fraud at the sales points of lottery tickets. This suspicion was also supported by other research findings. The investigations led to a great commotion. Headings rolled and the control procedures were adjusted to better protect the customer. The stores' ticket checking machines must now be viewable by customers, and make loud noises to indicate wins. Customers are now required to sign their names on their lottery tickets before redeeming them, to prevent switches.

4.2 Bombs over London in World War II

A famous application of the Poisson model is the statistical analysis of the distribution of hits of flying bombs (V-1 and V-2 missiles) in London during the second World War. The British authorities were anxious to know if these weapons could be accurately aimed at a particular target, or whether they were landing at random. If the missiles were in fact only randomly targeted, the British could simply disperse important installations to decrease the likelihood of their being hit. An area of 36 square kilometers in South London was divided into 576 small regions of 250 meter wide by 250 meter long, and the number of hits in each region was determined. There were 229 regions with zero hits, 211 regions with one hit, 93 regions with two hits, 35 regions with three hits, 7 regions with four hits, 1 region with five hits, and 0 regions with six or more hits. The 576 regions were struck by $229 \times 0 + 211 \times 1 + 93 \times 2 + 35 \times 3 + 7 \times 4 + 1 \times 5 = 535$ bombs and so the average number of hits per region was

$$\lambda = \frac{535}{576} = 0.9288.$$

If it can be made plausible that the number of hits per region closely follows a Poisson distribution with an expected value of 0.9288, then it can be safely concluded that the missiles landed at random, in view of properties of the two-dimensional Poisson process on the plane. You would expect

$$V_k = 576 \times e^{-0.9288} \frac{0.9228^k}{k!}$$

regions with exactly k hits for $k = 0, 1, \ldots$ if the number of hits per sector is approximately Poisson distributed with an expected value of 0.9288. Calculating V_k for $k = 0, 1, \ldots, 5$, you get

$V_0 = 227.5, V_1 = 211.3, V_2 = 98.1, V_3 = 30.4, V_4 = 7.1,$ and $V_5 = 1.3.$

You see that the observed relative frequencies 229, 211, 93, 35, 7, and 1 for the numbers of hits are each very close to these theoretical relative frequencies corresponding to a Poisson distribution (this can be formalized with the chi-square test from Section 3.8). It could be concluded that the distribution of hits in the South London area was much like the distribution of hits when a flying bomb was to fall on any of the equally sized regions with the same probability, independently of the other flying bombs. The statistical analysis convinced the British military that the bombs struck at random and had no advanced aiming ability.

4.3 Winning the lottery twice

The following item was reported in the February 14, 1986 edition of The New York Times: "A New Jersey woman wins the New Jersey State Lottery twice within a span of four months." She won the jackpot for the first time on October 23, 1985 in the 6/39 lottery. Then she won the jackpot in the new 6/42 lottery on February 13, 1986. In the r/s lottery, r different numbers are randomly drawn from the numbers 1 to s, and you win the jackpot if you have correctly predicted all six winning numbers. Lottery officials of New Jersey State Lottery declared that the probability of winning the jackpot twice in one lifetime is approximately one in 17.1 trillion. What do you think of this statement? The claim made in this statement is easily challenged. The officials' calculation proves correct only in the extremely farfetched case scenario of a given person submitting one ticket for the 6/39 lottery and one ticket for the 6/42 lottery just one time in his/her life. In this case, the probability of getting all six numbers right, both times, is equal to

$$\frac{1}{\binom{39}{6}} \times \frac{1}{\binom{42}{6}} = \frac{1}{1.71 \times 10^{13}}.$$

But the event of someone winning the jackpot twice is far from miraculous when you consider a very large number of people who play the lottery for many weeks. The explanation is the law of truly large numbers: *any event with a nonzero probability will eventually occur when it is given enough opportunity to occur.* Let's illustrate this lottery principle by considering the many 6/42 lotteries in the world that have two draws each week. Suppose that each of 50 million people fills in five tickets for each drawing. Then the probability of one of them winning the jackpot at least twice in the coming four years is close to 1. The calculation of this probability is based on the Poisson distribution, and goes as follows. The probability of winning the jackpot in a particular week when filling in five tickets is equal to $5/\binom{42}{6} = 0.531 \times 10^{-7}$. In view of the physical background of the Poisson distribution, the number of times that a given player will win the jackpot in the next $4 \times 52 \times 2 = 416$ drawings of a 6/42 lottery is modeled by a Poisson distribution whose expected value is

$\lambda_0 = 416 \times 9.531 \times 10^{-7} = 3.965 \times 10^{-4}$. For the next 416 drawings, this means that

P(a particular player will win the jackpot two or more times)
$= 1 - e^{-\lambda_0} - e^{-\lambda_0}\lambda_0 = 7.859 \times 10^{-8}$.

Next, using again the physical background of the Poisson distribution, we can conclude that the number of people under the 50 million mark, who win the jackpot two or more times in the coming four years is Poisson distributed with expected value

$$\lambda = 50\,000\,000 \times (7.859 \times 10^{-8}) = 3.93.$$

Thus, the probability that at some point in the coming four years at least one of the 50 million players will win the jackpot two or more times can be given as $1 - e^{-\lambda} = 0.980$. A probability very close to 1! A few simplifying assumptions are used to make this calculation, such as the players choose their six-number sequences randomly. This does not influence the conclusion that it may be expected once in a while, within a relatively short period of time, that *someone* will win the jackpot two times.

4.4 Santa Claus and a baby whisperer

In 1996, the James Randi Educational Foundation was founded by James Randi, a former top magician who fought and exposed mockery and pseudo-sciences. The goal of the foundation was to make the public and the media aware of the dangers associated with the performances of psychic mediums. James Randi offered a $1 million prize to anyone who could demonstrate psychic abilities. Obviously, this had to be demonstrated under verifiable test conditions, which would be agreed on beforehand. For example, someone like Uri Geller who claimed to be able to bend spoons without applying force could not bring his own spoons. Different mediums took up the challenge but nobody succeeded. The 'baby whisperer' Derek Ogilvie was one of the mediums who accepted the challenge. This medium claimed to be capable of extrasensory distant observations. He was allowed to choose a child with whom he thought he would have telepathic contact, and he was subjected to the following test. The medium was shown ten different toys that would be given to the child one after the other, in random order, out of sight of the medium. The child was taken to an isolated chamber, and each time the child received a toy, the medium was asked to say what toy it was. If the medium was right six or more times, he would win one million dollars. What is the probability of six or more correct answers?

The problem is in fact a variation of the Santa Claus problem: at a Christmas party, each one of a group of children brings a present, after which the children draw lots randomly to determine who gets which present. What is the probability that none of the children will wind up with their own present?[23] The Poisson heuristic will be used to get this probability. Suppose there are n children at the party and imagine that the children are numbered as $1, 2, \ldots, n$. The Santa Claus problem can be formulated within the framework of a sequence of n trials. In the ith trial, a lot is drawn by the child having number i. Let's say that a trial is successful if the child draws the lot for his/her own present. Then, the success probability of each trial has the same value $\frac{(n-1)!}{n!} = \frac{1}{n}$ (and so the order in which lots are drawn does not matter). Thus, the expected value of the number of successes is $n \times \frac{1}{n} = 1$, regardless of the value of n. The outcomes of the trials are not independent of each other, but the dependence is 'weak' if n is sufficiently large. The success probability $\frac{1}{n}$ is small for n large. Then, as noted before in Section 3.3, you can use the Poisson heuristic for the probability distribution of the total number of successes. This probability distribution is then approximated by a Poisson distribution with an expected value of 1. Thus, since a 'success' means that the child gets his/her own present, you get

$$P(\text{exactly } k \text{ children will get their own present}) \approx \frac{e^{-1}}{k!}, \quad 0 \leq k \leq n.$$

Numerical investigations reveal that this is a remarkably good approximation for $n \geq 10$ (the first seven decimals of the approximate values agree with the exact values already for n as large as 10). In particular, taking $k = 0$, the probability that none of the children

---

[23]The Santa Claus problem and its variations boil down to the following combinatorial problem. Take a random permutation of the integers $1, 2, \ldots, n$. What is the probability that none of the integers keeps its original position?

will get their own present is about $\frac{1}{e} = 0.36787\ldots$, or, about 36.8%, regardless of the number of children.

Going back to the ESP experiment with the medium, James Randi was in very little danger of having to cough up the loot. The probability of six or more correct answers for random guessing is practically equal to the Poisson probability $1 - \sum_{k=0}^{5} \frac{e^{-1}}{k!} = 0.0006$, or, about 0.06%. The medium perhaps thought, beforehand, that five correct guesses was the most likely outcome, and a sixth correct guess on top of that wasn't that improbable, so, why not go for it. In the test, he had only one correct answer.

4.5 Birthdays and 500 Oldsmobiles

In 1982, the organizers of the Quebec Super Lotto decided to use a fund of unclaimed winnings to purchase 500 Oldsmobiles, which would be raffled off as a bonus prize among the 2.4 million lottery subscribers in Canada. They did this by letting a computer randomly pick 500 times a number from the 2.4 million registration numbers of the subscribers. To the lottery officials' astonishment, they were contacted by one subscriber claiming to have won two Oldsmobiles. The lottery had neglected to program the computer not to choose the same registration number twice. The probability of a pre-specified subscriber winning the car two times is indeed astronomically small, but not so the case of the probability that, out of 2.4 million subscribers, there will be someone whose number appears at least twice in the list of 500 winning numbers. The latter event has a probability of about 5%. That is quite a small probability, but not a negligible probability. How can we calculate the probability of 5%? To do so, let's translate the lottery problem into a birthday problem on a planet with $d = 2\,400\,000$ days in the year and a randomly formed group of $m = 500$ aliens. What is the probability that two or more aliens share a birthday, assuming that each day is equally likely as birthday? This probability can be accurately approximated by

$$1 - e^{-\frac{1}{2}m(m-1)/d}.$$

This formula is easily obtained by the Poisson heuristic. There are $\binom{m}{2} = \frac{1}{2}m(m-1)$ different combinations of two aliens. Each combination has the same success probability of $\frac{1}{d}$ that the two aliens have a common birthday. In other words, you have a very large number of $\frac{1}{2}m(m-1)$ trials, each having the same tiny success probability $\frac{1}{d}$. The dependence between the trials is very weak and so the probability of no success is approximately equal to the Poisson probability $e^{-\lambda}$, where $\lambda = \frac{1}{2}m(m-1) \times \frac{1}{d}$ is the expected number of successful trials. This verifies the above approximation formula.

Identifying the Oldsmobiles with the aliens and the registration numbers with the birthdays, substitution of $d = 2\,400\,000$ for the probability of winning two or more Oldsmobiles by the same lottery subscriber. The approximation is very accurate and agrees with the exact value in the first five decimals. The exact formula for the probability of two or more matching birthdays is

$$1 - \frac{d(d-1)\cdots(d-m+1)}{d^m},$$

as can be seen by the same arguments as used for the classical birthday problem with 365 equally likely birthdays, see Section 1.2. In this birthday problem, 23 people suffice to have a fifty-fifty match probability under the assumption of equally likely birthdays. In reality, birthdays are not uniformly distributed throughout the year, but follow a seasonal pattern. However, for birth frequency variation as occurring in reality, the match probability is very insensitive to deviations from uniform birth rates. Empirical studies have been done that confirm this finding. For example, during the 2014 World Cup soccer championship, 32 national teams of 23 players each took part. It turned out that 18 of those teams had at least one double birthday (double birthdays for 15 teams at the 2018 World Cup soccer). The 2019 World Cup women's soccer had 24 teams of 23 players each and had 10 teams with at least one double birthday.

4.6 Cash Winfall lottery: a revenue model for stats geeks

This lottery was an obscure state lottery game that was gamed by sophisticated stats geeks. Some smart people had figured out how to get rich while everyone else funded their winnings. What made the Cash Winfall lottery unique is that the progressive jackpot could not grow past $2 million. Once it reached this level and there was no winner, the prizes for matching 3, 4, and 5 balls grew instead of the jackpot growing. And the smaller prizes were not parimutuel, i.e. winners did not have to share a fixed pot of money. So if the jackpot rose to \$2 million or more without a winner, the jackpot would 'roll-down' and instead be split among the players who had matched three, four, or five numbers. Lower-tier prizes were \$4000, \$150, or \$5 for matching five, four, or three numbers respectively, and those prizes were increased by a factor of five to ten if the jackpot reached \$2 million and was not won. In the lottery, six different numbers were drawn from the numbers 1 to 46.

Each week, the lottery published the estimated amount of the jackpot for that week's draw. Each time a roll-down draw approached, several syndicates bought a very large number of tickets. This was not too risky for them since the jackpot was seldom hit, and ordinary players barely bought more tickets as a roll-down draw approached. What can be said about the cash winnings of the syndicates? Let's say that one syndicate invested \$400 000 in 200 thousand lottery tickets of \$2 per ticket when a roll-down was expected. Under the assumption that those tickets are Quick Pick tickets whose ticket numbers are randomly generated by the lottery's computers, let's make some rough calculations for the case that the jackpot reached \$2 million and was not won. Let's take the conservative estimates

$$a_3 = \$27.50, \ a_4 = \$925, \ \text{and} \ a_5 = \$25\,000$$

for the payoff a_k on any ticket that matches exactly k of the six winning Winfall numbers. Denote by p_k the probability of a single ticket matching exactly k of the six winning numbers given that the jackpot was not won. Then p_k is $\binom{6}{k}\binom{40}{6-k}/\binom{46}{6}$ divided by $1 - 1/\binom{46}{6}$ and has the values

$$p_3 = 2.10957 \times 10^{-2}, \ p_4 = 1.24909 \times 10^{-3}, \ \text{and} \ p_5 = 2.56224 \times 10^{-5}.$$

Let's define the random variable X_k as the number of syndicate tickets that match exactly k of the six winning numbers for $k = 3, 4, 5$. In view of the physical background of the Poisson distribution, it is reasonable to approximate the distribution of X_k by a Poisson distribution with expected value $\lambda_k = 200\,000 \times p_k$. The numerical values of the λ_k are

$$\lambda_3 = 4219.149, \ \lambda_4 = 249.8180, \ \text{and} \ \lambda_5 = 5.12447.$$

This gives that the expected cash winnings of the syndicate can be estimated as

$$\lambda_3 a_3 + \lambda_4 a_4 + \lambda_5 a_5 = 475\,220 \text{ dollars.}$$

An expected profit of more than \$75 000, a healthy return of about 19%. What is an estimate for the probability that the syndicate does not a profit on its \$400 000 investment? In order to find this, we need the standard deviation of the cash winnings of the syndicate. The random variables X_3, X_4, and X_5 are nearly independent of each other. The variance of a Poisson variable X_k is equal to its mean, and so the standard deviation of cash winnings of the syndicate is approximately equal to

$$\sqrt{a_3^2 \lambda_3 + a_4^2 \lambda_4 + a_5^2 \lambda_5} = 58\,453 \text{ dollars.}$$

Next, note that the Poisson distribution can be accurately approximated by the normal distribution when the expected value of the distribution is not too small. Also, a linear combination of independent normally distributed random variables is again normally distributed. This gives that the distribution of the cash winnings $a_3 X_3 + a_4 X_4 + a_5 X_5$ can be approximated by a normal distribution with expected value \$475 220 and standard deviation \$58 453. Thus, the probability that the syndicate will lose money on its investment of \$400 000 can be estimated as

$$\Phi\left(\frac{400\,000 - 475\,220}{58\,453}\right) = 0.099.$$

Three syndicates, one of which was a group of MIT students, won millions of dollars by making clever use of the 'roll-down' character of the lottery, and in fact, profiting from a jackpot that had been amassed by other participants. By 2011, syndicate activity was getting a lot of negative publicity, which prompted lottery officials to adjust the rules, and ultimately to abandon the game altogether.

4.7 Coupon collecting

Suppose that a new brand of breakfast cereal is brought to the market. The producer has introduced a campaign offering one baseball card in each cereal box purchased. As baseball fan, you want to collect a complete set of baseball cards. What are the first two moments and the probability mass function of the number of cereal boxes you must buy in order to get all baseball cards? It is assumed that trading is not an option. The problem is an instance of the so-called *coupon collector's problem* that appears in many disguises: there are c different types of coupons labeled as $i = 1, \ldots, c$ and you will get one coupon with each purchase. This coupon will be coupon i with probability p_i, where $\sum_{i=1}^{c} p_i = 1$. How many purchases must be done in order to get a complete collection of coupons?

Case of equal probabilities

It is assumed that each coupon has the same probability of being collected, that is $p_i = \frac{1}{c}$ for $i = 1, \ldots, c$. Let the random variable N be the number of purchases needed to collect all c coupons. In the solutions of the Problems 2.48 and 2.54, it is shown that

$$E(N) = c \sum_{k=1}^{c} \frac{1}{k} \quad \text{and} \quad \sigma^2(N) = c^2 \sum_{k=1}^{c} \frac{1}{k^2} - c \sum_{k=1}^{c} \frac{1}{k}.$$

What about the probability distribution of the random variable N? It is not too difficult to compute the exact probability distribution of N. There are several methods. In Example 6.5 of Chapter 6, the powerful method of absorbing Markov chains is used. A practically useful approximation can be determined fairly easily with the help of the Poisson heuristic. Then a Poisson approximation to the probability that more than n purchases are needed to collect all coupons can be obtained by the following subtle argument. Take a fixed number of n purchases and imagine a series of c trials, where the ith trial refers to coupon i and this trial is said be *successful* if coupon i is *not* among the n purchases. The success probability of each trial is $\left(\frac{c-1}{c}\right)^n$. Thus, the expected value of the number of successful trials is $c(\frac{c-1}{c})^n$. Then, for c and n large enough, the distribution of the number of successful trials can be approximated by a Poisson distribution with expected value $c(\frac{c-1}{c})^n$. Thus, the probability of no successful trials can be approximated by $e^{-c(\frac{c-1}{c})^n}$. The probability

Table 2: Numerical results for the case of equal probabilities

n	150	200	250	300	350	400	500
app	0.9106	0.5850	0.2740	0.1101	0.0416	0.0153	0.0020
exa	0.9324	0.6017	0.2785	0.1109	0.0417	0.0154	0.0020

that not all c coupons are among the first n purchases is one minus the probability of no successful trials. Thus,

$$P(N > n) \approx 1 - e^{-c(\frac{c-1}{c})^n} \quad \text{for } n \geq c.$$

For $c = 50$ and several values of n, Table 2 gives the exact and approximate values of the probability that more than r purchases are needed to get a complete collection. Using the formulas

$$E(X) = \sum_{n=0}^{\infty} P(X > n) \quad \text{and} \quad E[X(X-1)] = \sum_{n=0}^{\infty} 2n\, P(X > n)$$

for a non-negative, integer-valued random variable X (see Problem 2.47), the approximate values 222.71 and 63.48 are obtained for $E(N)$ and $\sigma(N)$, where the exact values are 224.96 and 61.95. The Poisson heuristic gives excellent approximations.

Case of unequal probabilities

For the case of c coupons with unequal probabilities $p_1, \ldots, p_c$, the probability that more than n purchases are needed to get all c coupons can be approximated by

$$P(N > n) \approx 1 - (1 - e^{-np_1}) \times \cdots \times (1 - e^{-np_c}) \quad \text{for } n \geq c,$$

where $P(N > n) = 1$ for $0 \leq n \leq c - 1$. This approximation is motivated in the solution of Problem 3.24. It turns out that this is a very useful approximation. This is illustrated with the example of rolling two fair dice until each of the eleven possible sums has shown up. The probability p_k of getting the sum k in a single roll of two fair dice is $p_2 = p_{12} = \frac{1}{36}$, $p_3 = p_{11} = \frac{2}{36}$, $p_4 = p_{10} = \frac{3}{36}$, $p_5 = p_9 = \frac{4}{36}$, $p_6 = p_8 = \frac{5}{36}$, and $p_7 = \frac{16}{36}$. For several values of n, Table 3 gives the

Table 3: Numerical results for the case of unequal probabilities

n	25	50	75	100	125	150	200
app	0.9120	0.5244	0.2604	0.1277	0.0630	0.0304	0.0077
sim	0.9227	0.5168	0.2524	0.1225	0.0599	0.0294	0.0072

simulated and the approximate values of $P(N > n)$. The simulated values are based on ten million runs. The approximate values for $E(N)$ and $\sigma(N)$ are 61.72 and 36.82, where the exact values are equal to 61.22 and 35.98.

Multiple coupons

The approximation for $P(N > n)$ can be adapted for the coupon collector's problem in which you get d different coupons with each purchase. If the coupons are uniformly distributed among the packets with d coupons and d is much smaller than c, then numerical experiments indicate that

$$P(N > n) \approx 1 - \left(1 - e^{-n\frac{d}{c}}\right)^c$$

is a useful approximation. The rationale behind this approximation is that a specific coupon is contained in a packet of d coupons with probability $\frac{d}{c}$.[24] The famous Panini football sticker collection is an example of the coupon collector's problem with multiple coupons. The Italian firm Panini produces sticker albums for World Cup and Euro Cup soccer. The Euro Cup 2020 album had 678 stickers to collect, and you got a packet of six different stickers with each purchase. Under the assumption that all stickers are equally rare, a simulation study showed that you need 780, 878, and 993 packets of six stickers to have a complete album with probabilities of 50%, 75%, and 90%, respectively. In agreement with these simulation results, the approximation formula with $d = 6$ and $c = 678$ gives the values 779, 878,

---

[24]In the nonuniform case, this probability is approximately dp_k for coupon k when c is large, $d \ll c$ and *each p_j is small enough so that dp_j is a small probability for all j.* To explain this: the simplification that coupons are picked with replacement then has little impact and so the probability that coupon k is not in a packet of d coupons is approximately $(1 - p_k)^d \approx e^{-dp_k} \approx 1 - dp_k$.

and 991 for the smallest value of n such that $P(N > n)$ is less than 0.5, 0.25, and 0.10, respectively.

4.8 Benford's law

Benford's law, also called the Newcomb–Benford law, is a mathematical law about the leading digit number in real-world data. The law states that the first significant digit of numbers in various naturally occurring data sets does not always follow a uniform distribution, but rather the logarithmic probability distribution

$$\log_{10}\left(1 + \frac{1}{k}\right) \quad \text{for } k = 1, \ldots, 9.$$

In 1881, the famous astronomer Simon Newcomb (1835–1909) published a short article in which he observed that the initial pages of reference books containing logarithmic tables were far more worn and dog-eared than the later pages. He found that numbers beginning with a 1 were looked up more often than numbers beginning with 2, numbers beginning with 2 were looked up more often than numbers beginning with 3, etc. For digits 1 through 9, Newcomb found the relative frequencies to be

30.1%, 17.6%, 12.5%, 9.7%, 7.9%, 6.7%, 5.8%, 5.1%, 4.6%,

which is consistent with the mathematical formula $\log_{10}(1 + 1/k)$ for $k = 1, \ldots, 9$. This result was more or less forgotten until Frank Benford, an American physicist, published in 1938 an article in which he demonstrated empirically that the first nonzero digit in many types of data (lengths of rivers, metropolitan populations, universal constants in the fields of physics and chemistry, numbers appearing in front page newspaper articles, etc.) approximately follows a logarithmic distribution.

How to explain this law? Benford's law has the remarkable characteristic of being scale invariant: if a data set conforms to Benford's law, then it does so regardless of the physical unit in which the data are expressed. Whether river lengths are measured in kilometers or miles, or stock options are expressed in dollars or euros, it makes no difference for Benford's law. But this does not explain the frequent occurrence of the law. A satisfying mathematical explanation for Benford's law was a long time coming. In 1996, the American mathematician Ted Hill proved that if numbers are picked randomly from various randomly chosen sets of data ranging in orders of magnitude, the numbers from the combined sample approximately conform to Benford's law. This is a perfect description of what happens with numbers appearing in, e.g. front-page newspaper articles.

Although it may seem bizarre at first glance, the Benford's law phenomenon has important practical applications. In particular, Benford's law can be used for investigating financial data — income tax data, corporate expense data, corporate financial statements. Forensic accountants and taxing authorities use Benford's law to identify possible fraud in financial transactions. Many crucial bookkeeping items, from sales numbers to tax allowances, conform to Benford's law, and deviations from the law can be quickly identified using simple statistical controls. A deviation does not necessarily indicate fraud, but it does send up a red flag that will spur further research to determine whether or not there is a case of fraud. This application of Benford's law was successfully applied for the first time by a District Attorney in Brooklyn, New York. He was able to identify and obtain convictions in cases against seven fraudulent companies. In more recent years, the fraudulent Ponzi scheme of Bernard Madoff — the man behind the largest case of financial fraud in U.S. history — could have been stopped earlier if the tool of Benford's law had been used. Benford's law can also be used to identify fraud in macroeconomic data. Economists at the IMF have applied it to gather evidence in support of a hypothesis that countries sometimes manipulate their economic data to create strategic advantages, as Greece did in the time of the European debt crisis. This is a different kettle of fish altogether from the quaint application regarding dog-eared pages in old-fashioned books of logarithm tables. Nowadays, Benford's law has multiple statistical applications on a great many fronts. It is a little gem in data analysis.

4.9 What is casino credit worth?

In 1980, an Atlantic City casino extended a more or less unlimited credit line to gambling addict David Zarin. They only cut him off when his gambling debt passed the 3-million-dollar mark. Partly due to New Jersey state laws that provide a shield for gambling addicts, the casino in question had no legal recourse to collect the full amount of the debt; in fact, it was required to discharge the lion's share of the debt. But the story doesn't end there. Shortly after the court's decision was rendered, Zarin received a federal tax assessment claim from the Internal Revenue Service demanding tax payment on the sum of 3 million dollars, which had been defined as income. Zarin returned to court to fight this assessment, and won. His most important argument was that he had received no cash money from the casino, only casino chips with which to gamble. In the end, the court determined that Zarin did not owe taxes on the portion of the debt that had been discharged by the casino. This lawsuit generated much interest and has since been taken up into the canon of required case studies for law students studying in the United States.

In coming to its decision, the court neglected to ask this simple question: what monetary value can be assigned to a credit line of 3 million dollars in casino chips, that allows a player to gamble at a casino? First of all, it must be said that the odds of the player beating the casino are small. Still, the player does have a chance of beating the casino, of claiming a profit, and, after repaying the 3-million-dollar advance, of going home with a sum of money that was gained on the loan. The gambler's ruin formula enables us to quantify the monetary value of this loan.

Zarin's game at the casino was the tremendously popular game of craps. Craps is a dice game played with two dice. There are various betting options, but the most popular, by far, is the so-called 'pass-line' bet. It is not needed to get into the intricacies of the game or of pass-line betting procedure; suffice it to say that, using the pass-line bet, the probability that the player wins and the probability that the player loses have the values

$$p = \frac{244}{495} \quad \text{and} \quad q = \frac{251}{495},$$

respectively. When the player wins, the player gets a return of two times the amount staked; otherwise, the player loses the amount staked. This is precisely the situation of the classical gambler's ruin problem. In this problem, the gambler starts with a units of money, stakes one unit on each gamble, and then sees his bankroll increase by one unit with probability p and sees it decrease by one unit with probability $q = 1 - p$. The gambler stops when his bankroll reaches a predetermined sum of $a + b$ units of money, or when he has gone broke. Letting $P(a, b)$ be the probability that the gambler reaches his target of $a + b$ units of money without first having gone broke, the classical gambler's ruin formula is

$$P(a, b) = \frac{1 - (q/p)^a}{1 - (q/p)^{a+b}},$$

where $P(a, b)$ must be read as $a/(a + b)$ when $p = q = 0.5$. This formula will be proved in Section 6.3 of Chapter 6. The gambler's ruin formula can be used to show that, in David Zarin's case, it would not have been unreasonable to assign a value of 195 thousand dollars to his credit line of 3 million dollars. If a player wants to achieve the maximal probability of a predetermined winning sum in a casino game such as craps, then the best thing the player can do is to bet boldly, or rather, to stake the maximum allowable sum (or house limit) on each gamble. Intuition alone will tell us that betting the maximum exposes the player's bankroll to the casino's house edge for the shortest period of time. In Zarin's case, the casino had imposed a house limit of 15 thousand dollars for the pass-line bet in the game of craps. So, we may reasonably think that Zarin staked 15 thousand dollars on each gamble. In terms of the gambler's ruin formula then, 15 thousand dollars would be equal to one unit of money. We can further assume that Zarin's target goal was to increase his bankroll of $3\,000\,000/15\,000 = 200$ units of money by b units of money, having assigned a value to b beforehand. What is a reasonable choice for b when the casino gives a credit line of a units of money as starting bankroll? Part of the agreement is that the player will owe nothing to the casino if he goes broke, and the player will go home with a profit of b units of money if he increases his bankroll to $a + b$ units of money. The derivation of the best value of b uses the utility function $u(a, b)$, which is defined as the expected value of the sum with which the player will exit the casino. This utility function is given by

$$u(a, b) = b \times P(a, b) + 0 \times (1 - P(a, b)).$$

For a given bankroll a, a rational choice for b is that value for which $u(a, b)$, as a function of b, is maximal. This value of $u(a, b)$ for the maximizing b could be considered, by the court, to be the value of the credit advance of a units of money extended by the casino to the player. An insightful approximation can be given to the maximizing value of b, which we denote with b^*, and the corresponding value of the credit line. For the case of a sufficiently large bankroll a, it will be shown that

$$b^* \approx \frac{1}{\ln(q/p)} \quad \text{and} \quad u(a, b^*) \approx \frac{e^{-1}}{\ln(q/p)}.$$

Surprisingly, the value of the bankroll a is not relevant. These approximations can be derived by writing the gambler's ruin formula as

$$P(a, b) = \frac{(q/p)^{-a} - 1}{(q/p)^{-a} - (q/p)^b},$$

and noting that, for large a, the term $(q/p)^{-a}$ can be neglected when $q/p > 1$. Thus, $P(a, b) \approx \left(\frac{q}{p}\right)^{-b}$ and

$$u(a, b) \approx b \times \left(\frac{q}{p}\right)^{-b}.$$

Putting the derivative of $u(a, b)$ with respect to b equal to zero, the approximations for b^* and $u(a, b^*)$ follow after a little bit of algebra. An interesting result is that, for a sufficiently large bankroll a and a target amount of $b = b^*$, the probability of reaching the target is approximately equal to

$$e^{-1} \approx 0.3679,$$

regardless of the precise value of the bankroll a (note that $r^{1/\ln(r)} = e$ for any $r > 0$). If you apply these results to David Zarin's case, using the data $a = 200$, $p = \frac{244}{495}$ and $q = \frac{251}{495}$, then you find that

$$b^* \approx 35 \quad \text{and} \quad u(a, b^*) \approx 13.$$

This means that the value of the credit line extended by the casino is about 13 units of money. Each unit of money represents 15 thousand dollars. So, it can be concluded that the 3-million-dollar credit line extended to Zarin by the casino can be valued at about 195 thousand dollars. And that is the amount the American tax authorities would have been justified in taxing.

4.10 Devil's card game: a psychological test

This test was created for analyzing feelings of regret for a missed opportunity. The game underlying the test is played with 11 cards: an ace, a two, a three, a four, a five, a six, a seven, an eight, a nine, a 10, and a joker. Each card is worth its face value in points, while the ace counts for 1 point. To play the game, the cards are shuffled so that they are randomly arranged, and then they are turned over one at a time. You start with 0 points, and as you flip over each card your score increases by that card's points — as long as the joker has not shown up. The moment the joker appears, the game is over and your score is 0. The key is that you can stop any moment and walk away with a nonzero score. What strategy maximizes your expected number of points and how many points would you earn on average in the game?

Determining the optimal stopping rule is not difficult, unlike determining the maximum average score per game. The key to the solution is to see what the effect is on the current score when the player does not stop but draws one additional card from the remaining cards containing the joker. If your current score is p points and c cards have been drawn so far, then there are $10 - c$ non-joker cards left in the game with a total value of $55 - p$ points, where each of these cards has an average value of $(55 - p)/(10 - c)$ points. The next card you draw has a probability of $1/(11 - c)$ that it is the joker and a probability of $(10 - c)/(11 - c)$ that it is not the joker. Thus, if you continue, the expected increase of your current score of p points is

$$I = \frac{10 - c}{11 - c} \times \frac{55 - p}{10 - c} = \frac{55 - p}{11 - c},$$

whereas the expected decrease of your current score is

$$D = p \times \frac{1}{11 - c}.$$

The expected increase I is more than the expected decrease D if

$$\frac{55 - p}{11 - c} > \frac{p}{11 - c},$$

that is, $I > D$ if and only if $p < 27.5$. Intuitively, you would conclude that it is optimal to continue if the current score is below 28 and to stop otherwise. This "one-stage-look-ahead" rule is indeed the optimal stopping rule.[25] In optimal stopping theory, it is proved that the one-stage-look-ahead rule is optimal if the process has the property that it stays in the set of unfavorable states once it has entered this set. It is interesting to note that the optimal stopping rule does not depend on the number of cards already drawn, but depends only on the number of points gained so far.

Monte Carlo simulation can be used to find that the average score per game is 15.453 (with a standard deviation of 15.547 points) under the optimal stopping rule and to verify the remarkable fact that the probability of drawing the joker is exactly 50%. However, these results can also be analytically obtained without simulation. To that end, note that for the optimal stopping threshold of 28 points, you need at least 4 cards and at most 7 cards. Moreover, the final score cannot be more than 37 $(= 27 + 10)$ points. Then, using the fact that the probability of not getting the joker is $\binom{10}{c}/\binom{11}{c} = (11-c)/11$ when randomly drawing c cards from the 11 cards, you have

$$P(\text{the final score is } p \text{ points}) = \sum_{c=4}^{7} \frac{11 - c}{11} \times \frac{a_c(p)}{\binom{10}{c}} \text{ for } p = 28, \ldots, 37,$$

where $a_c(p)$ denotes the number of possible combinations of c distinct non-joker cards such that the sum of the values of the cards is equal to p and less than 28 if one of these c cards would be removed. The relevant values of p and c are small enough to determine the numbers $a_c(p)$ by enumeration. This leads to the values 0.5, 0.0952, 0.0826, 0.0751, 0.0643, 0.0536, 0.0442, 0.0356, 0.0249, 0.0162, and 0.0083 for

---

[25]If there are N cards with values $1, \ldots, N$ along with one joker card, the optimal threshold is $\frac{1}{2}S$, where $S = \frac{1}{2}N(N + 1)$ is the sum of the points of the non-joker cards. The optimal threshold becomes $\frac{1}{3}S$ if there are two jokers instead of one.

the probability of a final score of p points for $p = 0, 28, 29, \ldots, 37$. The expected value and the standard deviation of the final score follow from these probabilities. Since the probability of not picking the joker is 0.5, the end score has the expected value of $2 \times 15.453 = 30.906$ points under the condition that the joker has not been picked.

Problem 4.1. The thirteen cards of a particular suit are taken from a standard deck of 52 playing cards and are thoroughly shuffled. A dealer turns over the cards one at a time, calling out "ace, two, three, ..., king". A match occurs when the card turned over matches the rank called out by the dealer as he turns it over. What do you think are the chances of a match?

Problem 4.2. A blindfolded person is tasting ten different wines. Beforehand, he is informed of the names of the participating wineries, but is not told the order in which the ten wines will be served. The person may name each winery just once. During the taste-test, he succeeds in identifying five of the ten wineries correctly. Do you think this person is a wine connoisseur?

Problem 4.3. In the Massachusetts Numbers Game, lottery officials were confused when, as their lottery celebrated its second anniversary, they noticed that the same four-digit number had been drawn multiple times over the course of the 625 lottery draws that had taken place. They expected to run about 5 000 draws before they would encounter this phenomenon. Can you explain why it is almost certain that the same four-digit number will be drawn more than once in 625 draws?

Problem 4.4. Argue that $1 - e^{-\frac{3}{2}m(m-1)/d}$ approximates the probability that two or more aliens in the birthday problem from Section 4.5 have a birthday within one day from each other.

Problem 4.5. In the South African lottery, the numbers $5, \ldots, 10$ were drawn on Dec. 1, 2020. There was much fuss in the media and many people thought the lottery was a scam. With one hundred 6/42 lotteries with two draws per week, what are the chances of a draw with six consecutive numbers in some lottery in the next two years?

Chapter 5
Monte Carlo Simulation and Probability

Monte Carlo simulation is a natural partner for probability. It imitates a concrete probability situation on the computer. In this chapter you will see how simulation works and how you can simulate many probability problems with relatively simple tools.[26] You will notice that simulation is not a simple gimmick, but requires mathematical modeling and algorithmic thinking. The emphasis is on the modeling behind computer simulation, not on the programming itself.

5.1 Introduction

Monte Carlo simulation is a powerful probabilistic analysis tool, widely used in both engineering fields and non-engineering fields. It is named after the famous gambling hot spot, Monte Carlo, in the Principality of Monaco. Monte Carlo simulation was initially used to solve neutron diffusion problems in atomic bomb research at Los Alamos National Laboratory in 1944. From the time of its introduction during World War II, Monte Carlo simulation has remained one of the most-utilized mathematical tools in scientific practice. And in addition to that, it also functions as a very useful tool for adding an extra dimension to the teaching and learning of probability. It may help students gain a better understanding of probabilistic ideas and

---

[26]The law of large numbers is the mathematical basis for the application of computer simulation to solve probability problems. The probability of a given event in a chance experiment can be estimated by the relative frequency of occurrence of the event in a very large number of repetitions of the experiment.

Table 4: Simulation results for 100 000 coin tosses

n	$H_n - \frac{1}{2}n$	f_n	n	$H_n - \frac{1}{2}n$	f_n
10	1	0.6000	5 000	−9.0	0.4982
25	1.5	0.5600	7 500	11	0.5015
50	2	0.5400	10 000	24	0.5024
100	2	0.5200	15 000	40	0.5027
250	1	0.5040	20 000	91	0.5045
500	−2	0.4960	25 000	64	0.5026
1 000	10	0.5100	30 000	78	0.5026
2 500	12	0.5048	100 000	129	0.5013

to overcome common misconceptions about the nature of 'randomness'. As an example, a key concept such as the law of large numbers can be made to come alive before one's eyes by watching the results of many simulation trials. The nature of this law is best illustrated through the coin-toss experiment. The law of large numbers says that the percentage of tosses to come out heads will be as close to 50% as you can imagine, provided that the number of coin tosses is large enough. But how large is large enough? Experiments have shown that the relative frequency of heads may continue to deviate significantly from 0.5 after many tosses, though it tends to get closer and closer to 0.5 as the number of tosses gets larger and larger. The convergence to the value 0.5 typically occurs in a rather erratic way. The course of a game of chance, although eventually converging in an average sense, is a whimsical process. To illustrate this, a simulation run of 100 000 coin tosses was made. Table 4 summarizes the results of this particular simulation study; any other simulation experiment will produce different numbers. The statistic $H_n - \frac{1}{2}n$ gives the observed number of heads minus the expected number after n tosses and the statistic f_n gives the observed relative frequency of heads after n tosses. It is worthwhile to take a close look at the results in the table. You see that the realization of the relative frequency, f_n, indeed approaches the true value of the probability in a rather irregular manner and converges more slowly than most of us would expect intuitively. That is why you should be suspicious of the outcomes of simulation studies that consist of only a small number of simulation runs, see also Section 5.5.

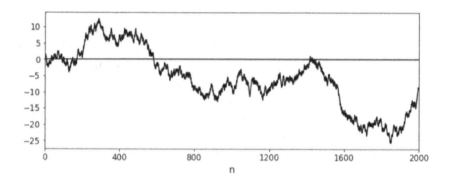

Figure 13: A random walk of 2 000 coin tosses.

The law of large numbers does not imply that the absolute difference between the actual number of heads and the expected number should oscillate close to zero. It is even typical for the coin-toss experiment that this difference has a tendency to become larger and larger and to grow proportionally with the square root of the number of tosses, whereby returns to 0 become rarer and rarer as the number of coin tosses gets larger and larger. This is illustrated in Figure 13 which displays a simulated realization of the random walk describing the actual number of heads minus the expected number over 2 000 coin tosses. The mathematical explanation of the growing oscillations displayed in Figure 13 is provided by the central limit theorem together with the square root law: the actual number of heads minus the expected number after n tosses is approximately normally distributed with expected value 0 and standard deviation $\frac{1}{2}\sqrt{n}$ for large n. The coin-toss experiment is full of surprises that clash with intuitive thinking. Unexpectedly long sequences of either heads or tails can occur ('local clusters' of heads or tails are absorbed in the average). If you don't believe this, convince yourselves with simulation. Simulation can reveal interesting and surprising patterns.

Monte Carlo simulation is not only a very useful tool for helping students to gain a better understanding of probabilistic ideas and to overcome common misconceptions about the nature of "randomness", but also enables you to get quick answers to specific probability problems or to check analytical solutions. For example, what is the probability that any two adjacent letters are different when the eleven letters of the word Mississippi are put in random order? Seemingly a simple probability problem, but it turns out that this combinatorial probability problem is difficult to solve analytically. In combinatorial probability, it is often beforehand not clear whether a probability problem easily allows for an analytical solution. Many probability problems are too difficult or too time-consuming to solve exactly, while a simulation program is easily written. Monte Carlo simulation can also be used to settle disagreement on the correct answer to a particular probability problem. It is easy to make mistakes in probability, so checking answers is important. Take the famous Monty Hall problem. This probability puzzle raised a lot of discussion about its solution. Paul Erdös, a world famous mathematician, remained unconvinced about the correct solution of the problem until he was shown a computer simulation confirming the correct result. In the Monty Hall problem, a contestant in a TV game show must choose between three doors. An expensive car is behind one of the three doors, and gag prizes are behind the other two. He chooses a door randomly, appealing to Lady Luck. Then the host opens one of the other two doors and, as promised beforehand, takes a door that conceals a gag prize. With two doors remaining unopened, the host now asks the contestant whether he wants to remain with his choice of door, or whether he wishes to switch to the other remaining door. What should the contestant do? Simulating this game is a convincing approach to show that it is better to switch, which gives a win probability of $\frac{2}{3}$.

5.2 Simulation tools

Simple tools often suffice for the simulation of probability problems. This section discusses first the concept of random generator. Next several useful simulation tools are presented. These tools include methods to generate a random point inside a bounded region and a random permutation of a finite set of objects. The simulations tools will be illustrated in the next section.

5.2.1 Random number generators

In the simulation of probability models, access to random numbers is of crucial importance. A *random number generator*, as it is called, is indispensable. A random number generator produces random numbers between 0 and 1 (excluding the values 0 and 1). For a 'truly' random number generator, it is as if fate falls on a number between 0 and 1 by pure coincidence. A random number between 0 and 1 is characterized by the property that the probability of the number falling in a sub-interval of $(0,1)$ is the same for each interval of the same length and is equal to the length of the interval. A truly random number can take on any possible value between 0 and 1. A random number from $(0,1)$ enables you to simulate, for example, the outcome of a single toss of a fair coin without actually tossing the coin: if the generated random number is between 0 and 0.5 (the probability of this is 0.5), then the outcome of the toss is heads; otherwise, the outcome is tails. Producing random numbers is not as easily accomplished as it seems, especially when they must be generated quickly, efficiently, and in massive amounts. Even for simple simulation experiments, the required amount of random numbers runs quickly into the hundreds of thousands or higher.[27] Generating a very large amount of random numbers on a one-time only basis, and storing them up in a computer memory, is practically infeasible. But there is a solution to this kind of practical hurdle that is as handsome as it is practical.

Instead of generating *truly* random numbers, a computer can generate so-called *pseudo random numbers*, and it achieves this through a nonrandom procedure. This idea comes from the famous Hungarian-American mathematician John von Neumann (1903–1957) who made very important contributions not only to mathematics but also to physics and computer science. The procedure for a pseudo random number generator is iterative by nature and is determined by a suitably chosen function f. Starting with a seed number z_0, numbers $z_1, z_2, \ldots$ are successively generated by $z_1 = f(z_0)$, $z_2 = f(z_1)$, and

---

[27]In earlier times creative methods were sometimes used to generate random numbers. Around 1920 crime syndicates in New York City's Harlem used the last five digits of the daily published U.S. treasure balance of the American Treasury to generate the winning numbers for their illegal 'Treasury Lottery'.

so on.[28] The function f is referred to as a pseudo random number generator and it must be chosen such that the sequence $\{z_i\}$ is statistically indistinguishable from a sequence of truly random numbers. The output of function f must be able to stand up to a great many statistical tests for 'randomness'.

The first pseudo random number generators were the so-called multiplicative congruential generators. Starting with a positive integer z_0, the z_i are generated by $z_i = az_{i-1}$ (modulo m) for $i = 1, 2, \ldots$, where a and m are carefully chosen positive integers, e.g. $a = 16\,807$ and $m = 2^{31} - 1$. Then the sequence $\{z_i\}$ repeats itself after $m - 1$ steps, and so m is the cycle length. The number z_i determines the random number u_i by $u_i = \frac{z_i}{m}$.

The newest pseudo random number generators do not use the multiplicative congruential scheme. In fact, they do not involve multiplications or divisions at all. These generators are very fast, have incredibly long periods before they repeat the same sequence of random numbers, and provide high-quality pseudo random numbers. In software tools, you will find not only the so-called Christopher Columbus generator with a cycle length of about 2^{1492} (at ten million pseudo random numbers per second, it will take more than 10^{434} years before the sequence of numbers will repeat!), but you will also find the Mersenne twister generator with a cycle length of $2^{19937} - 1$. This generator would probably take longer to cycle than the entire future existence of humanity. It has passed numerous tests for randomness, including tests for uniformity of high-dimensional strings of numbers. The modern generators are needed in Monte Carlo simulations requiring huge masses of pseudo random numbers, as is the case in applications in physics and financial engineering.

In the sequel, we omit the additive 'pseudo' and simply speak of random number and random number generator.

5.2.2 Simulating from a finite range

How do you choose randomly a number between two given numbers a and b with $a < b$? To do so, you first use the random number

---

[28]Pseudo random number generators enable you to reproduce a sequence of random numbers by using the same seed. This is practically useful when you want to compare random systems under the same experimental conditions.

generator to get a random number u between 0 and 1. Next, you find a random number x between a and b as

$$x = a + (b - a)u.$$

How do you choose randomly an integer from the integers $1, 2, \ldots, M$? To do so, you first use the random number generator to get a random number u between 0 and 1. Then a random integer k is

$$k = 1 + \text{int}(M \times u).$$

The function $\text{int}(x)$ rounds the number x to the nearest integer k that is smaller than or equal to x. That is,

$$\text{int}(x) = k \quad \text{if } k \leq x < k + 1$$

for an integer k. More generally, a random integer k from the integers $a, a + 1, \ldots, b$ is obtained as

$$k = a + \text{int}\big((b - a + 1) \times u\big).$$

This can be used to simulate the outcome of a roll of a fair die.

The procedure for selecting a random integer can be used to simulate an outcome in a chance experiment in which each outcome is equally likely. A more general case is that of a chance experiment in which each possible outcome has a probability that is a multiple of $\frac{1}{r}$ for some integer r. Then, you can use the ingenious *array method* to simulate an outcome. In this method, you only need to generate one random integer from the integers 1 to r. To explain the method, consider a chance experiment with three possible outcomes O_1, O_2, and O_3 with probabilities $p_1 = 0.50$, $p_2 = 0.15$, and $p_3 = 0.35$, respectively. Then $r = 100$ and you form the array A[i] for $i = 1, \ldots, 100$ with $\text{A}[1] = \cdots = \text{A}[50] = 1$, $\text{A}[51] = \cdots = \text{A}[65] = 2$ and $\text{A}[66] = \cdots = \text{A}[100] = 3$. You generate a random number u between 0 and 1. Next, you calculate $k = 1 + \text{int}(100 \times u)$, being a random integer from $1, \ldots, 100$. Then $\text{A}[m]$ gives you the random observation from the probability mass function. For example, suppose $u = 0.63044\ldots$ has been generated. Then, $m = 64$ with $\text{A}[64] = 2$. This gives the random outcome O_2.

5.2.3 Simulating a random permutation

Suppose you have 10 people and 10 labels numbered as 1 to 10. How to assign the labels at random such that each person gets assigned a different label? This can be done by making a random permutation of the integers $1, \ldots, 10$ and assigning the labels according to the random order in the permutation. An algorithm for generating a random permutation is useful for many probability problems. A simple and elegant algorithm can be given for generating a random permutation of $(1, 2, \ldots, n)$. The idea of the algorithm is first to randomly choose one of the integers $1, \ldots, n$ and to place that integer in position n. Next you randomly choose one of the remaining $n - 1$ integers and place it in position $n - 1$, etc.

Algorithm for random permutation

1. Initialize $t := n$ and $a[j] := j$ for $j = 1, \ldots, n$.

2. Generate a random number u between 0 and 1.

3. Set $k := 1 + \text{int}(t \times u)$ (random integer from the integers $1, \ldots, t$). Interchange the current values of $a[k]$ and $a[t]$.

4. $t := t - 1$. If $t > 1$, return to step 2; otherwise, stop with the random permutation $(a[1], \ldots, a[n])$.

As an illustration, the algorithm is used to construct a random permutation of the integers 1, 2, 3, and 4.

Iteration 1. $t := 4$. If the generated random number $u = 0.71397\ldots$, then $k = 1 + \text{int}(4 \times 0.71397\ldots) = 3$. Interchanging the elements of the positions $k = 3$ and $t = 4$ in $(1, 2, 3, 4)$ gives $(1, 2, 4, 3)$.
Iteration 2. $t := 3$. If the generated random number $u = 0.10514\ldots$, then $k = 1 + \text{int}(3 \times 0.10514\ldots) = 1$. Interchanging the elements of the positions $k = 1$ and $t = 3$ in $(1, 2, 4, 3)$ gives $(4, 2, 1, 3)$.
Iteration 3. $t := 2$. If the generated random number $u = 0.05982\ldots$, then $k = 1 + \text{int}(2 \times 0.05982\ldots) = 1$. Interchanging the elements of the positions $k = 1$ and $t = 2$ in $(4, 2, 1, 3)$ gives $(2, 4, 1, 3)$.
Iteration 4. $t := 1$. The algorithm stops with the random permutation $(2, 4, 1, 3)$.

Verify yourselves that the algorithm carries literally over to the construction of a random permutation of a finite sequence of objects in which some objects appear multiple times.

The algorithm can also be used to simulate a random subset of integers. For example, how to simulate a draw of the lotto 6/45 in which six distinct numbers are randomly drawn from the number 1 to 45? This can be done by using the algorithm with $n = 45$ and performing only the first 6 iterations until the positions $45, 44, \ldots, 40$ are filled. Then $a[45], \ldots, a[40]$ give the six numbers for the lottery draw.

5.2.4 Hit-and-miss method

Let's first discuss how to choose a random point inside a rectangle.

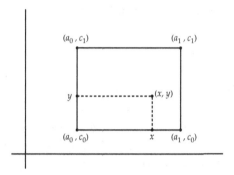

Figure 14: Simulating a random point inside a rectangle.

Let (a_0, c_0), (a_1, c_0), (a_0, c_1), and (a_1, c_1) be the four corner points of the rectangle, see Figure 14. You first use the random number generator to get two random numbers u_1 and u_2 from $(0, 1)$. Then the random point (x, y) inside the rectangle is

$$x = a_0 + (a_1 - a_0)u_1 \quad \text{and} \quad y = c_0 + (c_1 - c_0)u_2.$$

How do you generate a random point inside a circle? To do this, you face the complicating factor that the coordinates of a random point inside a circle cannot be generated independently of each other. Any point (x, y) inside a circle with radius r and the origin $(0, 0)$ as center must satisfy $x^2 + y^2 < r^2$. A tempting procedure is to generate first a random number $x = a$ from the interval $(-r, r)$ and to generate next

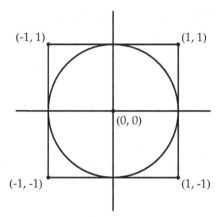

Figure 15: Simulating a random point inside the circle.

a random number y from the interval $(-\sqrt{r^2 - a^2}, \sqrt{r^2 - a^2})$. This procedure, however, violates the requirement that the probability of the random point falling into a sub-region should be the same for any two sub-regions having the same area.

A simple but powerful method to generate a random point inside the circle is the *hit-and-miss method*. The idea of this method is to take a rectangle that envelopes the bounded region and to generate random points inside the rectangle until a point is obtained that falls inside the circle. This simple approach can be used to generate a random point inside any bounded region in the plane. As an illustration, take the unit circle with radius 1 and the origin $(0,0)$ as center. The circle is clamped into the square with the corner points $(-1, -1)$, $(1, -1)$, $(-1, 1)$, and $(1, 1)$, see Figure 15. A random point (x, y) inside this square is found by generating two random numbers u_1 and u_2 from $(0, 1)$ and taking x and y as

$$x = -1 + 2 \times u_1 \ \text{ and } \ y = -1 + 2 \times u_2.$$

Next, you test whether
$$x^2 + y^2 < 1.$$

If this is the case, you have found a random point (x, y) inside the unit circle; otherwise, you repeat the procedure. On average, you have to generate $\frac{4}{\pi} = 1.273$ random points inside the square until you get a random point inside the circle. To see this, note that $\frac{4}{\pi}$ is the ratio of the area of the square and the area of the unit circle.

The idea of the hit-and-miss method is generally applicable. It is often used to find the area of a bounded region in the two-dimensional plane or a higher-dimensional space by enveloping the region by a rectangle or in a higher-dimensional cube. These areas are typically represented by multiple integrals. A classic example of the use of Monte Carlo simulation to compute such integrals goes back to the analysis of neutron diffusion problems in the atomic bomb research at the Los Alamos National Laboratory in 1944. The physicists Nicholas Metropolis and Stanislaw Ulam had to compute multiple integrals representing the volume of a 20-dimensional region, and they devised the hit-and-miss method for that purpose.

5.2.5 Rejection sampling

Rejection sampling is an extension of the hit-and-miss method. It underlies many specialized algorithms to simulate from specific probability distributions. The method will be given for the case of a continuously distributed random variable with a general probability density $f(x)$, but the method also applies to a discrete random variable. Rejection sampling is used when it is difficult to sample directly from the probability density $f(x)$. Instead of sampling directly from $f(x)$, rejection sampling uses an envelope density $g(x)$ from which sampling is easier. The proposal density $g(x)$ must satisfy

$$f(x) \leq cg(x) \quad \text{for all } x$$

for some constant c. Note that c must be at least 1 since both $f(x)$ and $g(x)$ integrate to 1. Rejection sampling goes as follows:

Step 1. Simulate a candidate x from $g(x)$ and a random number $u \in (0, 1)$.

Step 2. If $u \leq \frac{f(x)}{cg(x)}$, then accept x as a sample from $f(x)$; otherwise, repeat step 1.

The rejection sampling method is illustrated in Figure 16. The particular sample shown in the figure will be rejected. Since the average

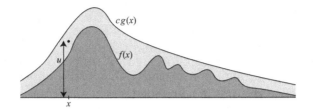

Figure 16: Rejection sampling.

number of iterations needed to obtain an accepted draw can be shown to be equal c, the method is only attractive when c is not too large. Rejection sampling is a very useful method for simulating from one-dimensional densities. However, in high dimensions the method is impractical. Apart from the difficulty of finding an envelope density, the bounding constant c will be typically very large.

Let's illustrate how the rejection sampling method can be used to sample from the standard normal random variable Z. To do so, observe that a random sample from $X = |Z|$ yields a random sample from Z by letting Z be equally likely to be either X or $-X$. The probability density of X is

$$f(x) = \frac{2}{\sqrt{2\pi}} e^{-\frac{1}{2}x^2} \quad \text{for } 0 < x < \infty.$$

As proposal density $g(x)$, the exponential density e^{-x} for $0 < x < \infty$ is taken. It is easy to simulate from the exponential density by applying the inverse transform method. This method will be described below. The smallest constant c such that $f(x) \leq cg(x)$ for all $x > 0$ is found as

$$\max_{x>0} \frac{f(x)}{g(x)} = \max_{x>0} \sqrt{\frac{2}{\pi}} e^{x-0.5x^2}.$$

The maximum is achieved for $x = 1$ and so $c = \sqrt{\frac{2e}{\pi}} \approx 1.315$.

Inverse transform method

How to sample a random observation from an exponential distributed random variable? This can be done by the inverse transform method.

This method is the preferred method for a random variable X having a cumulative probability distribution function $F(x) = P(X \leq x)$ that is strictly increasing in x and has an easily computed inverse function F^{-1}. The method is based on the fact that the random variable $F^{-1}(U)$ has the same probability distribution as X if the random variable U is uniformly distributed on $(0, 1)$. This fact is easy to verify: $P(F^{-1}(U) \leq x) = P(U \leq F(x)) = F(x)$, since $P(U \leq u) = u$ for $0 \leq u \leq 1$. The inverse transform method goes as follows:

Step 1. Generate a random number $u \in (0, 1)$.
Step 2. Calculate $x = F^{-1}(u)$. Then x is a random observation of the random variable X.

For the case that X has an exponential density $\mu e^{-\mu x}$ for $x > 0$, the cumulative distribution function $F(x) = 1 - e^{-\mu x}$ for $x \geq 0$. This function is strictly increasing and has the inverse function

$$F^{-1}(u) = -\frac{1}{\mu} \ln(1 - u),$$

as follows by solving the equation $1 - e^{-\mu x} = u$. Since $1 - U$ is also uniformly distributed on $(0, 1)$, you can also use $-\frac{1}{\mu} \ln(u)$ as random observation. Simulation of the Poisson process becomes easy, using the inverse transform method for the exponential density.

5.3 Probability applications of simulation

Monte Carlo simulation is a powerful tool for getting numerical answers to probability problems, which are otherwise too difficult for an analytical solution. Various examples will be given to illustrate this. Simulation can also be used as a sanity check for an analytical solution or as a validation tool for an approximate solution. Randomization of deterministic algorithms is another area in which random numbers can be used to improve the average-case performance of the algorithm. This will be illustrated with the quick-sort algorithm.

5.3.1 Geometric probability problems

Geometric probability problems constitute a class of probability problems that often seem very simple but are sometimes very difficult to solve analytically. Take the problem of finding the expected value of the distance between two random points inside the unit square (sides with length 1) and the expected value of two random points inside the unit circle (radius 1). The analytical derivation of these expected values requires advanced integral calculus and leads to the following exact results:

$$\frac{1}{15}\left[2 + \sqrt{2} + 5\ln(1 + \sqrt{2})\right] = 0.5214 \quad \text{and} \quad \frac{128}{45\pi} = 0.9054.$$

It is a piece of cake to estimate the expected values by computer simulation. How does the simulation program look like? Perform a very large number of simulation runs. In each simulation run, two random points (x_1, y_1) and (x_2, y_2) are generated inside the unit square or the unit circle, see subsection 5.2.4. In each run, the distance between the two points is calculated by Pythagoras as

$$\sqrt{(x_1 - x_2)^2 + (y_1 - y_2)^2}.$$

Then, the average of the distances found in the simulation runs is calculated. By the law of large numbers, the average gives an estimate for the expected value of the distance between two random points. Many simulation runs are needed to get accurate estimates. The question of how many runs should be done will be addressed in the next section. In one million simulation runs, the estimates 0.5213 and 0.9053 were obtained for the expected values of the distances between two random point inside the unit square and between two random points inside the unit circle. The computing times for the one million simulation runs are a matter of seconds on a computer.

5.3.2 Almost-birthday problem

In the classic birthday problem the question is what the probability is that two or more people share a birthday in a randomly formed group of people. This problem was analytically solved in Section 1.2. It was easy to find the answer. The problem becomes much more difficult when the question is what the probability is that two or more people have a birthday within one day from each other. This is the *almost-birthday problem*. However, the simulation program for the almost-birthday problem is just as simple as the simulation program for the birthday problem. An outline of the simulation programs is as follows. The starting point is a randomly formed group of m people (no twins), where each day is equally likely as birthday for any person. For ease, it is assumed that the year has 365 days (February 29 is excluded). For each of the two birthday problems, a very large number of simulation runs is performed. In each simulation run, m random integers $g_1, \ldots, g_m$ are generated, where the random integer g_i represents the birthday of the ith person. Each of these integers is randomly chosen from the integers $1, \ldots, 365$, see subsection 5.2.2. In each simulation run for the classic birthday problem you test whether there are distinct indices i and j such that

$$|g_i - g_j| = 0,$$

while in each simulation run for the almost-birthday problem you test whether there are distinct indices i and j such that

$$|g_i - g_j| \leq 1 \text{ or } |g_i - g_j| = 364.$$

You find an estimate for the sought probability by dividing the number of simulation runs for which the test criterion is satisfied by the total number of runs. As you see, the simulation program for the almost-birthday problem is just as simple as that for the classic birthday problem.

5.3.3 Consecutive numbers in lottery

What is the probability of getting two or more consecutive numbers when six distinct numbers are randomly drawn from the numbers 1 to 45 in the lotto 6/45? The exact value of this probability is

$$1 - \frac{\binom{40}{6}}{\binom{45}{6}} = 0.5287,$$

but the argument to get this result is not simple. However this probability, which is surprisingly large, can be quickly and easily obtained by computer simulation. In each simulation run, you get the six lottery numbers by applying six iterations of the algorithm from subsection 5.2.3 and taking the six integers in the array elements $a[45], a[44], \ldots, a[40]$. Next, you test whether there are two or more consecutive numbers among these six numbers. This is easily done by checking whether $a[i] - a[j]$ is 1 or -1 for some i and j with $40 \leq i, j \leq 45$. The desired probability is estimated by dividing the number of simulations runs for which the test criterion is satisfied by the total number of simulation runs. One million simulation runs resulted in the estimate 0.5289. The simulation program needs only a minor modification to simulate the probability of getting three or more consecutive numbers. Simulation leads to the estimate 0.056 for this probability.

5.3.4 Mississippi problem

An amusing but very difficult combinatorial probability problem is the Mississippi problem. What is the probability that any two adjacent letters are different in a random permutation of the eleven letters of the word Mississippi? A simulation model can be constructed by identifying the letter m with the number 1, the letter i with the number 2, the letter s with the number 3, and the letter p with the number 4. In each simulation run, a random permutation of the sequence $(1, 2, 3, 3, 2, 3, 3, 2, 4, 4, 2)$ is constructed by using an obvious modification of the permutation algorithm from subsection 5.2.3: the initialization of the algorithm now becomes $a[1] = 1$, $a[2] = 2$, ..., $a[10] = 4$, $a[11] = 2$. To test whether any two adjacent numbers are different in the resulting random permutation $(a[1], a[2], \ldots, a[11])$, you check whether $a[i + 1] - a[i] \neq 0$ for $i = 1, \ldots, 10$. The estimate 0.058 was obtained for the sought probability after 100 000 simulation runs.

5.3.5 Venice-53 lottery: what's in a number?

Misconceptions over the way that truly random sequences behave fall under the heading gambler's fallacy. This refers to the gambler who believes that, if a certain event occurs less often than average within a given period of time, it will occur more often than average during the next period. This misconception is persistent in the roulette game. The gambler's fallacy is also behind the Venice-53 hysteria in the national Italian lottery when the number 53 remained elusive for many months in the bi-weekly Venice lottery draw. Monte Carlo simulation can help convince people that such a remarkable happening is less coincidental than it seems.

In the Italian Regional Lottery, there is a bi-weekly draw in each of ten Italian cities, including Venice. Each draw, in each of the ten cities, consists of five numbers picked from the numbers 1 through 90. While the number 53 had fallen repeatedly in other cities, it did not come up at all in Venice, in any of the 182 draws occurring in the period from May, 2003, to February, 2005. During this period, more than 3.5 billion euro was bet on the number 53, entire family fortunes risked. In a frenzy that even lottery-mad Italy has rarely seen, some 53 addicts ran up debts, went bankrupt, and lost their homes to the bailiffs. In the month of January, 2005, alone, 672 million euro were staked on the number 53. Several professors of probability theory made appearances on Italian television to alert people to the fact that lottery balls have no memory, an attempt to stave off irresponsible betting behavior. All in vain. Many Italians held firmly to the belief that the number 53 was about to fall, and they continued to bet large sums. Some tragic events occurred during that January, 2005, direct consequences of the enormous amounts of money bet and lost on the number 53 in the Venice lottery. After 182 successive draws resulting in no number 53, that number finally made its appearance in the February 9, 2005 draw, thus bringing an end to the gambling frenzy that had held sway over Italy for such a long time. It is estimated that the lottery paid out about 600 million euro to those that had placed bets on the number 53 on that day. This is a lot of money, but it is nothing compared to the amount taken in by the lottery during the Venice-53 gambling craze. It is only a question of time before Venice-53 history repeats itself. Using Monte Carlo simulation, it can be demonstrated that there is a high probability that this will happen within 10 years, or 25 years of the event. A simulation model can be easily set up, using the procedure from subsection 5.2.5 for a random draw of five different numbers from numbers 1 to 90. Simulation shows that there is a probability of about 50% that in a period of 10 years there is some number that will not appear in some of the ten Italian city lotteries in a window of 182 or more consecutive draws, while this probability is about 91%

for a period of 25 years. It is to be feared that these hard facts will have little power to prevent future outbreaks of lottery madness.

A heuristic solution approach can also be given, which can be validated by simulation. Take a specific city (say, Venice) and a particular number (say, 53). The probability Q_n that there will be a window of 182 consecutive draws in which number 53 does not appear in the next n draws of the Venice lottery can be exactly calculated, see Problem 6.9 in Chapter 6. This probability has the values $Q_n = 0.0007754$ for $n = 1\,040$ and $Q_n = 0.0027013$ for $n = 2\,600$. The numbers drawn from 1 to 90 in the lottery are not independent of each other, but the dependence is weak enough to justify the approximation $(1-Q_n)^{90}$ for the probability that in the next n draws of the Venice lottery there will *not* be some number that remains absent during 182 consecutive draws. The lottery takes place in 10 cities. Thus, using again the unsurpassed complement rule, $1-(1-Q_n)^{10\times90}$ gives approximately the probability that there will be some number not appearing in one or more of the 10 Italian city-lotteries during 182 or more consecutive draws within the next n draws. Simulation experiments show that this approximation is remarkably accurate. It has the values 0.5025 for $n = 1\,040$ (period of 10 years) and 0.9124 for $n = 2\,600$ (period of 25 years).

5.3.6 Kruskal's count and another card game

A fascinating card game, or magic trick if you like, goes by the name of Kruskal's count. The game is not only fun, but Kruskal's principle has also useful applications in computer science and cryptography. The card game goes like this: a magician invites a spectator to thoroughly shuffle a deck of cards. Then the magician lays out the cards, face up, in one long row (or in a grid). Each card has a number value: aces have the value 1, the face cards (king, queen, jack) have the value 5, and the number cards have the value of the number on the card. The spectator is asked to think of a secret number from one to 10. The magician explains that the spot corresponding to that number, in the row of cards, is the spectator's first 'key card', and that the value of this key card determines the distance, in steps, to the next key card. If the secret number chosen by the player is 7, then the 7th card in the row of cards will be the spectator's first key card. If the 7th card is a 4, then the 11th card in the row is the new key card. If the 11th card is a jack, the 16th card in the row is the new key card, etc. The spectator counts in silence until reaching a key card with a value that doesn't permit continuing on because the deck has been exhausted and there aren't enough cards left. This ultimate key card is called the spectator's 'final card'. The magician then predicts which card is the final card. And more often than not, the magician will be right! So, what's the trick? It is astonishingly simple. Just as the spectator does, the magician also chooses a secret number between 1 and 10, and starting with this initial key card, silently counts through the row of cards just as the spectator does. It appears that there is a high probability that the two paths will meet at some point in their 'walk' through the sequence of cards, and from that point on, the paths remain the same.[29] An exact formula for the magician's success probability in Kruskal's count is not known, but the success probability can be easily found by Monte Carlo simulation. If the spectator and the magician each, blindly and independently of one another, choose a number between 1 and 10 for the starting state, then Monte Carlo simulation with one million runs gives an estimate of about 84% for the probability of the magician correctly 'predicting' which card is the spectator's final card. The success probability of the magician increases to about 97% when a double deck of 104 playing cards is used.[30]

The Humble-Nishyama card game

This is another interesting card game whose solution requires Monte Carlo simulation. You play against an opponent using an ordinary deck of 52 cards consisting of 26 black (B) cards and 26 red (R) cards, thoroughly shuffled. Before play starts, each player chooses a three-card-code sequence of red and black. For example, your opponent chooses BBR and you choose RBR. The cards are laid on the table, face up, one at a time. Each time that one of the two

---

[29]The success of the trick is based on path coupling in random walks.

[30]The approximation $1 - \left(1 - \frac{1}{a^2}\right)^N$ can be given for the probability of a correct guess of the magician, where N is the number of cards and a is the average value of a card (a has the value $\frac{70}{13}$ in the above cases).

chosen sequences of red and black appears, the player of this sequence gets one point. The cards that were laid down are removed and the game continues with the remaining cards. The player who collects the most points is the winner, with a tie being declared if both players have the same number of points. Your opponent is first to choose a sequence. The 64 000 dollar question is this: how can you choose, in response to the sequence chosen by your opponent, in such a way as to give you the maximum probability of winning the game? The counter-strategy is simple and renders you a surprisingly high win probability. The first element in your counter-move should be the opposite of the second element in the sequence chosen by your opponent. The last two elements in your counter-move should be the same as the first two elements in the sequence of your opponent. Your opponent chooses first a three-card-code sequence of red and black. The possible choices for your opponent are BBB, BBR, BRB, BRR, RRR, RRB, RBR, RBB, which leaves you to parry with the counter-moves RBB, RBB, BBR, BBR, BRR, BRR, RRB, RRB. Then, using Monte Carlo simulation with one-million runs for each case, the values 0.995, 0.935, 0.801, 0.883, 0.995, 0.935, 0.801, 0.883 are found for your probability of winning the card game. The probability of the card game ending in a tie has the values 0.004, 0.038, 0.083, 0.065, 0.004, 0.038, 0.083, and 0.065, respectively.

5.3.7 Randomized quick-sort algorithm

By adding randomness into a deterministic algorithm, the average-case performance of the algorithm can sometimes be improved. A nice example is the quick-sort algorithm that sorts the elements of a large array of data. The basic idea of the algorithm for sorting a given array of distinct elements is:

1. Pick an element of the array as pivot.
2. Compare each element of the array with the pivot and generate two sub-arrays A_1 and A_2, where A_1 contains the elements that are smaller than the pivot and A_2 contains the elements that are greater than the pivot.
3. Recursively sort array A_1 and array A_2.

In the basic version of quick-sort, the choice of the pivot is deterministic, usually the first element of the array is taken. For example, suppose the elements of the following array must be sorted:

$$[9, 5, 12, 15, 7, 4, 8, 11].$$

Picking 9 as pivot, you get

$$[5, 7, 4, 8], 9, [12, 15, 11],$$

where the pivot 9 is on its final position. In the worst-case scenario of the basic version of quick-sort, the number of comparisons needed is $\sum_{k=1}^{n} k = \frac{1}{2}n(n-1)$. A better criterion for the performance of the algorithm is the average number of comparisons needed, but in the deterministic algorithm this number depends very much on the input values. By adding randomization to the algorithm, the average number of comparisons needed can be guaranteed to be on the order of $n \ln(n)$, no matter how the input values are distributed. In randomized quick-sort, the pivot element is chosen at random from the n elements of the array. Let the random variable C_n be the number of comparisons needed by randomized quick-sort on an array of n elements. To find $E(C_n)$, condition on the rank of the pivot. If the randomly chosen pivot is the jth smallest among the n elements, then the conditional distribution of C_n is equal to the unconditional distribution of $n - 1 + C_{j-1} + C_{n-j}$. Let $\mu_n = E(C_n)$. Then, by the law of conditional expectation,

$$\mu_n = \sum_{j=1}^{n} E\left(n - 1 + C_{j-1} + C_{n-j}\right)\frac{1}{n} = \sum_{j=1}^{n} \left(n - 1 + \mu_{j-1} + \mu_{n-j}\right)\frac{1}{n}.$$

Since $\mu_0 = 0$, this expression can be rewritten as

$$\mu_n = n - 1 + \frac{2}{n} \sum_{k=1}^{n-1} \mu_k \quad \text{for } n \geq 1.$$

Without details of the derivation, the solution of this recurrence equation is

$$\mu_n = \sum_{i=1}^{n-1} 2\left(\frac{1}{2} + \cdots + \frac{1}{n - i + 1}\right) \quad \text{for } n \geq 1.$$

Thus, by the asymptotic expansion of the partial sum of the harmonic series in Section 1.2, you get that $E(C_n)$ is of the order $2n \ln(n)$ for larger values of n.

5.4 Bootstrap method in data analysis

The bootstrap method differs from traditional statistical methods by letting the data speak for themselves, using the number-breaking power of modern-day computers. The method is used in situations that you have a representative random sample from a population and other samples from the population cannot be drawn. Assuming that the data are random observations that are independent of each other and are representative of the population, the bootstrap method takes the sample data that a study obtains, and then resamples it over and over to create many simulated samples. Each bootstrap sample has the same size as the original sample and is created by sampling at random from the original data *with replacement*. Why is it called bootstrapping method? The term "bootstrapping" originated with a phrase in use in the 18th and 19th century: "to pull oneself up by one's bootstraps" (the tale of Baron von Münchhausen who pulled himself up by the bootstraps out of a swamp). The bootstrap is not a magic technique that provides a panacea for all statistical inference problems, but it has the power to substitute tedious and often impossible analytical derivations with Monte Carlo calculations. Bootstrapping is a way to estimate the variation of a statistic based on the original data. Let's give two applications of bootstrapping.

Comparing two groups of data

In order to test a new skin infection remedy, twenty healthy volunteers are infected with the corresponding ailment. They are then randomly split up into two groups of equal size: a remedy group and a placebo group. The study being a randomized double-blind study, the volunteers are not aware of which group they are assigned to, and the doctors do not have this information either. Each volunteer undergoes daily examinations until the malady is cured. In the remedy group, the values for the number of days required until all patients are cured are 7, 9, 9, 11, 12, 14, 15, 15, 15, and 17. In the placebo group, the values for the number of days until all patients are cured are 9, 11, 11, 11, 12, 15, 17, 18, 18, and 20. In order to test whether the remedy helps or not, the difference in the total number of days until cured between the placebo group and the remedy group is taken as test statistic. For the original sample, this statistic has the value $142 - 124 = 18$. In order to make a statistical statement of whether the remedy works or not, you combine the $m = 10$ data points of the placebo group and the $n = 10$ data points of the remedy group and you assume that it does not matter whether or not the remedy is used. Under this so-called null-hypothesis, the twenty data points are seen as twenty independent observations from a same unknown probability distribution. In each bootstrap run, you draw at random $m = 10$ data points for the placebo group and $n = 10$ data points for the remedy group from the original combined data set without replacement (thus, each of the twenty original data points in the combined set is equally likely to be chosen as new data point). Then the difference between the sum of the $m = 10$ values drawn and the sum of the $n = 10$ values drawn is calculated. Using 10 000 bootstrap runs, it was found that the proportion of runs in which this difference is 18 or more is 0.135. A probability of 0.135 is not small enough to reject the null-hypothesis. The conclusion is that more investigations should be done before any definitive conclusion can be reached about the remedy's effectiveness.

Predicting election results

In some polling method, voters are not asked to choose a favorite party, but instead they are asked to indicate how likely they are to vote for each party. Suppose that there are three parties A, B, and C. Let's assume that a representative group of 1 000 voters is polled. A probability distribution (p_{iA}, p_{iB}, p_{iC}) with $p_{iA} + p_{iB} + p_{iC} = 1$ describes the voting behavior of respondent i for $i = 1, \ldots, 1 000$. That is, p_{iX} is the probability that respondent i will vote for party X on election day. The voting probability distributions of the 1 000 voters can be summarized in 8 groups that are given in Table 5: the vote of each of 230 people will go to parties A, B, and C with probabilities 0.20, 0.80, and 0, the vote of each of 140 people will go to parties A, B, and C with probabilities 0.65, 0.35, and 0, and so on. Each person votes independently of the other people.

Table 5: Voting probabilities

No. of voters	$(p_{iA}, \ p_{iB}, \ p_{iC})$
230	(0.20, 0.80, 0)
140	(0.65, 0.35, 0)
60	(0.70, 0.30, 0)
120	(0.45, 0.55, 0)
70	(0.90, 0.10, 0)
40	(0.95, 0, 0.05)
130	(0.60, 0.35, 0.05)
210	(0.20, 0.55, 0.25)

How do we calculate probabilities such as the probability that party A will become the largest party and the probability that parties A and C together will get the majority of the votes? This can be done by the bootstrap method. In each simulation run, 230 draws are done from the distribution $(0.2, 0.8, 0)$, 140 random draws from $(0.65, 0.35, 0)$, and so on. A Monte Carlo simulation with 100 bootstrap runs each with 1 000 new data points leads to the following bootstrap estimates with their 95% confidence intervals:

$P(\text{party } A \text{ becomes the largest party}) = 0.120 \, (\pm 0.002)$
$P(\text{party } B \text{ becomes the largest party}) = 0.872 \, (\pm 0.002)$
$P(\text{parties } A \text{ and } C \text{ get the majority of the votes}) = 0.851 \, (\pm 0.002)$.

5.5 Statistical analysis of simulation output

It is never possible to achieve perfect accuracy through simulation. All you can measure is how likely the estimate is to be correct. It is important to have a probabilistic judgment about the accuracy of the point estimate. Such a judgment is provided by the concept of confidence interval. You will see that, if you want to achieve one more decimal digit of precision in the estimate, you have to increase the number of simulation runs with a factor of about one hundred. In other words, the probabilistic error bound decreases as the reciprocal of the square root of the number of simulation runs.

Suppose that you want to estimate the unknown probability p of a particular event E. To that end, n simulation runs are done. Let the indicator variable X_i be 1 if event E occurs in the ith simulation run and X_i be 0 otherwise. Then,

$$\hat{p} = \frac{1}{n} \sum_{i=1}^{n} X_i$$

is an estimator for the true value of p. The accuracy of this estimator is expressed by the 95% confidence interval

$$\left(\hat{p} - 1.96 \frac{\sqrt{\hat{p}(1-\hat{p})}}{\sqrt{n}}, \ \hat{p} + 1.96 \frac{\sqrt{\hat{p}(1-\hat{p})}}{\sqrt{n}} \right).$$

The 95% confidence interval should be interpreted as follows: any such interval to be constructed by simulation will cover the true value of p with a probability of about 95% when n is large. In other words, in about 95 out of 100 cases, the confidence interval covers the true value of p. Each simulation study gives an other confidence interval!

The effect of n on the term $\sqrt{\hat{p}(1-\hat{p})}$ fades away quickly if n gets larger. This means that the width of the confidence interval is nearly proportional to $1/\sqrt{n}$ for n sufficiently large. This conclusion leads to a practically important rule of thumb:

to reduce the width of a confidence interval by a factor of two, about four times as many observations are needed.

This is a very useful rule for simulation purposes. Let's illustrate this with the almost-birthday problem with a group of 20 people. For the probability that two or more people will have their birthdays within one day of each other, a simulation with 25 000 runs results in the probability estimate of 0.8003 with (0.7953, 0.8052) as 95% confidence interval, whereas 100 000 simulation runs result in an estimate of 0.8054 with (0.8029, 0.8078) as 95% confidence interval. The confidence interval has indeed been narrowed by a factor of 2.

How can you construct a confidence interval if the simulation study is set up to estimate an unknown expected value of some random variable X rather than an unknown probability? Letting $X_1, \ldots, X_n$ represent the observations for X resulting from n independent simulation runs, then estimators for $\mu = E(X)$ and $\sigma = \sigma(X)$ are

$$\hat{\mu} = \frac{1}{n} \sum_{k=1}^{n} X_k \quad \text{and} \quad \hat{\sigma} = \sqrt{\frac{1}{n} \sum_{k=1}^{n} (X_k - \hat{\mu})^2}.$$

The 95% confidence interval for the unknown $\mu = E(X)$ is

$$\left(\hat{\mu} - 1.96 \frac{\hat{\sigma}}{\sqrt{n}}, \ \hat{\mu} + 1.96 \frac{\hat{\sigma}}{\sqrt{n}} \right).$$

The statistic $\hat{\sigma}/\sqrt{n}$, which is the estimated standard deviation of the *sample mean* $\hat{\mu}$, is usually called the *standard error* of the sample mean. Any 95% confidence interval to be constructed by simulation will cover the true value of μ with a probability of about 95% when n is large. In other words, the confidence level specifies the long-run percentage of intervals containing the true value of μ. It is really important to reflect on this probability statement. The parameter μ does not vary: it is fixed, but unknown. The random element in the probability statement are the two random endpoints of the confidence interval around μ.

The foregoing will be illustrated with an inventory problem known as the newsboy problem. In this problem, a newsboy decides at the beginning of each day how many newspapers he will purchase for resale. Let's assume that the daily demand for newspapers is uniformly distributed between 150 and 250 papers. Demand on any given day is independent of demand on any other day. Assume that at the beginning of each day the newsboy purchases 217 newspapers. The purchase price per newspaper is one dollar. The resale price per newspaper is two dollars; the agency will buy back any unsold newspapers for fifty cents apiece. One hundred simulation experiments were performed, where in each experiment the sales over $n = 2\,000$ days were simulated and a confidence interval was constructed for

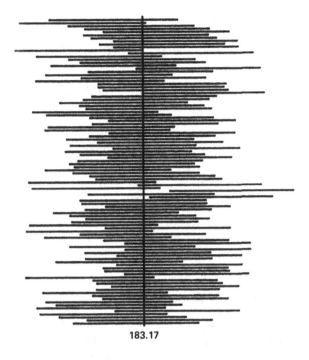

183.17

Figure 17: One hundred confidence intervals.

the expected net profit on a given day. Figure 17 displays the resulting one hundred 95% confidence intervals for the expected value of the net profit on any given day. The exact value of this expected value is $\mu = 183.17$, as can be analytically shown. It is instructive to have a look at the figure. Indeed, in approximately 95 of the 100 cases, the true value of μ is contained within the confidence interval.

To conclude, let's sketch how the 95% confidence interval is obtained from the central limit theorem. By this theorem, $(\frac{1}{n}\sum_{i=1}^{n} X_i - \mu)/(\sigma/\sqrt{n})$ has approximately a standard normal distribution for large n. This result remains true when σ is replaced by its estimator $\hat{\sigma}$. The standard normal distribution has 95% of its mass between the percentiles $z_{0.025} = -1.96$ and $z_{0.975} = 1.96$. Thus, for large n,

$$P\left(-1.96 \leq \left(\frac{1}{n}\sum_{i=1}^{n} X_i - \mu\right)/(\hat{\sigma}/\sqrt{n}) \leq 1.96\right) \approx 0.95.$$

This can be rewritten as

$$P\Big(\frac{1}{n}\sum_{i=1}^{n}X_i - 1.96\frac{\hat{\sigma}}{\sqrt{n}} \le \mu \le \frac{1}{n}\sum_{i=1}^{n}X_i + 1.96\frac{\hat{\sigma}}{\sqrt{n}}\Big) \approx 0.95.$$

Voila, the 95% confidence interval $(\hat{\mu} - 1.96\frac{\hat{\sigma}}{\sqrt{n}}, \hat{\mu} + 1.96\frac{\hat{\sigma}}{\sqrt{n}})$. The estimated standard deviation $\hat{\sigma}$ does not vary much if n gets larger, and so the width of the confidence interval is proportional to $\frac{1}{\sqrt{n}}$ for larger values of n, implying that you need about four times as many simulation runs in order to reduce the width of the interval by a factor of two. To illustrate, let the random variable X be the distance between two randomly chosen point in the unit square, see subsection 5.3.1. One hundred thousand simulation runs gives the estimate 0.5220 for $\mu = E(X)$ with the 95% confidence interval $(0.5204, 0.5235)$, and four hundred thousand simulation runs gives the estimate 0.5212 with the 95% confidence interval $(0.5205, 0.5220)$.

5.5.1 Variance reduction through importance sampling

The estimation of very small probabilities with standard simulation generally requires a great deal of computing time; the number of runs needed to find a confidence interval with a given relative precision is inversely proportional to the value of the probability that is to be estimated. Importance sampling is useful in such situations.

Let $f(x)$ be the probability density of a random variable X and suppose you want to estimate

$$E[a(X)] = \int a(x)f(x)\,dx$$

for some function $a(x)$. Importance sampling is a useful simulation method for doing this when the variance of the random variable $a(X)$ is very large. The idea is to sample from another probability density $g(y)$ to achieve variance reduction, using the observation that

$$E[a(X)] = \int a(x)f(x)\,dx = \int \Big[a(y)\frac{f(y)}{g(y)}\Big]g(y)\,dy = E_Y\Big[a(Y)\frac{f(Y)}{g(Y)}\Big],$$

where Y is an appropriately chosen random variable with probability density $g(y)$. The random variable

$$Z = a(Y)\frac{f(Y)}{g(Y)}$$

is the importance sampling estimator for $E(a(X))$. This is the basis of importance sampling. For large n, let $y_1, \ldots, y_n$ be independent samples from $g(y)$. Then, by the strong law of large numbers,

$$E[a(X)] \approx \frac{1}{n} \sum_{i=1}^{n} w_i\, a(y_i) \quad \text{with } w_i = \frac{f(y_i)}{g(y_i)}.$$

To estimate a small probability of the form $P(X \in A)$, apply the above with $a(x) = 1$ for $x \in A$ and $a(x) = 0$ for $x \notin A$ so that $E[a(X)] = P(X \in A)$. In this particular case, the second moment of the importance sampling estimator Z is

$$E(Z^2) = \int_A \left(\frac{f(y)}{g(y)}\right)^2 g(y)\, dy = \int_A \frac{f(y}{g(y} f(y)\, dy.$$

Comparing the second moment of the importance sampling estimator Z with the second moment $E[a(X)^2] = \int_A f(x)\, dx$ of the standard estimator $a(X)$, you see that variance reduction is achieved if $f(y)/g(y) < 1$ for all $y \in A$. In other words, the new probability density $g(x)$ must have a greater mass on the set A than the old probability density $f(x)$. The choice of $g(x)$ is a subtle matter and is problem dependent.

As an illustration, consider the problem of estimating the probability that a random walk with a negative drift will exceed a value $b > 0$ before dropping below a value $a < 0$ for given a and b. Suppose that the step sizes $X_1, X_2, \ldots$ of the random walk are independent random variables each having an $N(\mu, \sigma^2)$ density with $\mu < 0$. Let θ be defined as the probability that the random walk starting at the origin exceeds b before it drops below a. How can importance sampling be used to estimate the probability θ more effectively than with standard simulation? To do so, consider a random walk with positive drift, where each of the independent step sizes $Y_1, Y_2, \ldots$ has the $N(-\mu, \sigma^2)$ density $g(x)$. This density puts more mass on (b, ∞) than the original density $f(x)$ so that the event of exceeding the level b before dropping below the level a will occur more often in the modified random walk than in the original random walk. Now, for any $k \geq 1$, define the function $a(x_1, \ldots, x_k)$ as

$$a(x_1, \ldots, x_k) = \begin{cases} 1 & \text{for } x_1 + \cdots + x_k > b, \\ 0 & \text{otherwise.} \end{cases}$$

Define the random variable N as the smallest n for which $Y_1 + \ldots + Y_n$ is either less than a or greater than b. Then, the importance sampling estimator $Z = a(Y_1, \ldots, Y_N) \prod_{i=1}^{N} \frac{f(Y_i)}{g(Y_i)}$ is

$$Z = a(Y_1 + \cdots + Y_N) \, e^{2\mu(Y_1 + \cdots + Y_N)/\sigma^2}.$$

The expected value of this estimator is the sought probability θ. The variance of the importance sampling estimator is less than the variance of the standard estimator. For example, taking $\mu = -1$, $\sigma = 1$, $a = -4$, and $b = 3$ and doing $100\,000$ simulation runs, the 95% confidence interval 0.00078 (± 0.00017) is found for θ under standard simulation and the 95% confidence interval 0.000793 $(\pm 4.3 \times 10^{-5})$ under importance-sampling simulation. A considerable reduction in the width of the confidence interval.

Simulation modeling problems

In each of the following modeling problems you are asked to set up a mathematical model for a simulation program for the problem in question. It is fun and instructive to write a simulation program, using a programming language such as Python or R.

Problem 5.1. Set up a simulation model in order to estimate the probability that the equation $Ax^2 + Bx + C = 0$ has two real roots if A, B, and C are randomly chosen numbers from $(-1, 1)$. Do the same if A, B, and C are randomly chosen nonzero integers between $-1\,000$ and $1\,000$.

Problem 5.2. You draw random numbers from $(0, 1)$ until the sum of the picked numbers is larger than 1. Set up a simulation model to estimate the expected number of picks needed.

Problem 5.3. A dealer draws random numbers from $(0, 1)$ until the sum exceeds a predefined value a with $0 < a < 1$. The dealer has to beat the sum a without exceeding 1. Set up a simulation model to estimate the probability of the dealer winning the game.

Problem 5.4. You randomly pick two points on a stick. At these points the stick is broken into three pieces. Set up a simulation model to estimate the probability that a triangle can be formed with the pieces.

Problem 5.5. Three points are randomly chosen inside the unit circle. Set up a simulation model in order to estimate the probability that the center of the circle is contained in the triangle formed by the three random points.

Problem 5.6. You randomly choose two points inside a circle. Set up a simulation model to estimate the probability that these two random points and the center of the circle form an obtuse triangle. Do the same for a sphere in which two random points are chosen.

Problem 5.7. Set up a simulation model to estimate the expected value of the area of the triangle that is formed by three randomly chosen points inside the unit square. Do the same for three randomly chosen points inside the unit circle. *Hint:* the area of a triangle with sides of lengths a, b, and c is $\sqrt{s(s-a)(s-b)(s-c)}$, where $s = \frac{1}{2}(a+b+c)$.

Problem 5.8. Let $f(x)$ be a positive function on a finite interval (a,b) such that $0 \le f(x) \le M$ for $a \le x \le b$. How can you use simulation to estimate the integral $\int_a^b f(x)\,dx$?

Problem 5.9. Set up a simulation model to estimate the expected value of the distance between two randomly chosen points inside the unit cube. Do the same for the unit sphere.

Problem 5.10. A die is rolled until either each of the six possible outcomes has appeared or one of these outcomes has appeared six times. Set up a simulation model to estimate the probability that the first event will occur before the second event.

Problem 5.11. Seven students live in a same house. Set up a simulation model to estimate the probability that two or more of them have their birthdays within one week of each other.

Problem 5.12. The eight teams that have reached the quarter-finals of the Champions League soccer consist of two British teams, two German teams, two Italian teams, and two Spanish teams. Set up a simulation model to estimate the probability that no two teams from the same country will be paired in the quarter-finals draw if the eight teams are paired randomly.

Problem 5.13. Sixteen teams remain in a soccer tournament. A drawing of lots will determine which eight matches will be played. Before the drawing takes place, it is possible to place bets with bookmakers over the outcome of the drawing. Set up a simulation model to estimate the probability mass function of the number of correctly predicted matches.

Problem 5.14. Consider Problem 3.2 again. Set up a simulation model to estimate the probability of drawing the same pupil's name three or more times.

Problem 5.15. Consider Problem 2.41 again. Set up a model for simulating the probability histogram of the number of unpicked chickens.

Problem 5.16. In European roulette, a popular betting system is the d'Alembert system. You play this system with a starting bankroll of $500 and you bet a multiple of the unit stake of $5 each time. Starting with a bet of one unit, your bet is increased by one unit after a loss and decreased by one unit after a win. Each bet is either on red or black. The game continues until 100 bets have been made or you have lost your entire bankroll. Set up a simulation model to estimate the probability histogram of your end capital and the casino's average profit per unit staked.

Problem 5.17. Consider the devil's card game from Section 4.10 again. Suppose you stop as soon as your score is s or more. Set up a simulation model to estimate the probability of an end score of zero and the expected value of your end score as function of s.

Problem 5.18. You roll two fair dice until each of the 11 possible values of the sum of the roll has appeared. Set up a simulation model to estimate the expected number of rolls needed.

Problem 5.19. Set up a simulation model to estimate the expected number of draws needed in lottery 6/45 to get each of the 45 lottery numbers.

Problem 5.20. Five boys and five girls are lined up in random order. Set up a simulation model to estimate the probability that none of the girls is surrounded by two boys.

Problem 5.21. You have 10 cards, where each card is labeled with one of the numbers1 to 5. For each of these numbers there are two cards with that number. You deal the cards out randomly to five people so that each person gets two cards. Set up a simulation model to estimate the probability that two people have the same hand.

Problem 5.22. On two consecutive days, the same n people sit in random order at a round dining table. Set up a simulation model to validate the approximation $e^{-2}\left(1 - \frac{4}{n} + \frac{20}{3n^3}\right)$ for the probability that no two people sit next to each other at both dinners.

Problem 5.23. Set up a simulation model to estimate the probability of getting either five or more consecutive heads or five or more consecutive tails, or both, in 25 tosses of a fair coin.[31]

Problem 5.24. Consider Problem 3.9 again. You now bet each time the same fraction $\frac{2}{7}$ of your current bankroll (this is the so-called Kelly strategy that maximizes the growth rate of your bankroll on the long run). Set up a simulation model to estimate the probability distribution of your end capital.

---

[31]Surprisingly long runs can occur in coin tossing. A rule of thumb says that the probability mass function of the longest run of either heads or tails, or both, in n tosses of a fair coin is strongly concentrated around $\log_2(\frac{1}{2}n) + 1$ for larger values of n. A true story in this regard is the following. At the end of the 19th century, the Le Monaco newspaper regularly published the results of roulette spins in the casino of Monte Carlo. The famous statistician Karl Pearson (1857–1936) studied these data to test his theories. He observed that red and black came up a similar number of times, but also noticed that lengths of runs of either reds or blacks were much shorter than he would have expected. What caused that? Well, it turned out that the journalists at Le Monaco just made up the roulette results at the bar of the casino. They didn't think anybody would notice.

Problem 5.25. You have three fair dice and you initially roll all three dice. Any die that falls on a six is 'banked' and set aside. You continue to roll all the unbanked dice again until all the dice are banked. Set up a simulation model to find the probability mass function of the number of rolls required to bank all three dice.

Problem 5.26. The Benford game is a new casino game. First, the casino's computer generates a random number u and calculates the largest four-digit number below $10^3 \times 10^u$. The player cannot see this number until he has chosen his own four-digit number. The player's number is then multiplied by the casino's number. The player wins if the product number begins with a 4, 5, 6, 7, 8, or 9; otherwise, the casino wins. Set up a simulation model to investigate the claim that the casino wins with a probability of about 60%, regardless of the number chosen by the player.

Problem 5.27. A Christmas party is held for 10 persons. Each person brings a gift to the party for a gift exchange. The gifts are numbered 1 to 10 and each person knows the number of their own gift. Cards with the numbers 1 to 10 are put in a hat. The party goers consecutively pull a card out of the hat at random. If a person pulls out a card corresponding to the number of their own gift, then the card is put back in the hat, and that person draws another card. Each time the next person to take a card is chosen at random. Set up a simulation model to estimate the probability that the last person gets stuck with his or her own gift.

Problem 5.28. A famous TV show is The Tonight Show with Jimmy Fallon. In this show Jimmy plays the Egg Russian Roulette game with a guest of the show. The guest is always a celebrity from sports or film. The guest and Jimmy take turns picking an egg from a carton and smashing it on their heads. The carton contains a dozen eggs, four of which are raw and the rest are boiled. Neither Jimmy nor the guest knows which eggs are raw and which are boiled. The first person who has cracked two raw eggs on their head loses the game. The guest is the first to choose an egg. Set up a simulation model to estimate the probability that the guest will lose the game.

Problem 5.29. A radio station presents the following game every day. A number is drawn at random from the numbers 1 to 100. Listeners can call the station and guess the number drawn. If a caller guesses correctly, the caller gets a prize and the game is over; if not, the station informs the listeners whether the guess is too high or too low. The next caller then randomly guesses a number that has not yet been excluded. Set up a simulation model to estimate the probability distribution of the number of trials required.

Problem 5.30. There are ten chocolates in a Christmas tree, two of which are white and the other eight are dark. You take chocolates from the tree, one at a time and at random, and eat them until you pick a chocolate of the other color. You hang this chocolate back and start again with the leftover chocolates. Set up a simulation model to estimate the probability that the last chocolate you eat is white.

Problem 5.31. A retired gentleman considers participation in an investment fund. An adviser shows him that with a fixed yearly return of 14%, which was realized in the last few years, he could withdraw $15 098 from the fund at the end of each of the coming 20 years. This is music to the ears of the retired gentleman, and he decides to invest $100 000. However, the yearly return fluctuates. If the return was $r\%$ for the previous year, then for the coming year the return will remain at $r\%$ with a probability of 0.5, will change to $0.8r\%$ with a probability of 0.25, and will change to $1.2r\%$ with a probability of 0.25. Set up a simulation model to estimate the probability distribution of the capital left after 15 years.

Problem 5.32. In the show game Big Wheel, the wheel contains 20 sections showing cash values from $0.05 to $1.00 in 5-cent increments. Three contestants take turns spinning the wheel once or twice with the goal of getting as close to $1.00 as possible without going over it. If a participant decides to do a second spin, it must be done immediately after the first spin and the value of the second spin is added to that of the first spin. The winner is the player who comes closest to $1.00 without going over it. In case of a tie, the tied players draw lots to determine the winner. Set up a simulation model to find the optimal winning strategy for each player.

Problem 5.33. An opaque bowl contains 11 envelopes in the colors red and blue. You are told that there are four envelopes of one color each containing $100 and seven empty envelopes of the other color, but you cannot see the envelopes in the bowl. The envelopes are taken out of the bowl, one by one and in random order. Every time an envelope is taken out, you have to decide whether or not to open this envelope. Once you have opened an envelope, you get the money in that envelope (if any), and the process stops. Your stopping rule is to open the envelope drawn as soon as four or more envelopes of each color have been taken out of the bowl. Set up a simulation model to estimate your probability of winning $100.

Problem 5.34. In the first 1240 draws of the UK National Lottery, a record gap of length 72 appeared on 4th November 2000. The number 17 did not appear for 72 consecutive draws in this 6/49 lottery. Set up a simulation model to find the probability that some number will not appear during 72 or more consecutive draws in the next 1 240 draws of the lottery.

Problem 5.35. An alarm is triggered by events occurring according to a Poisson process at a rate of 0.5 per hour. If three or more events occur in a 15–minute time interval, the alarm sounds. Set up a simulation model to find the probability that the alarm will sound in a 15–minute time interval somewhere in a 24–hour period.

Problem 5.36. A beer company brings a new beer with the brand name Babarras to the market and prints one of the letters of this brand name underneath each bottle cap. Each of the letters A, B, R, and S must be collected a certain number of times in order to get a nice beer glass. The quota for the letters A, B, R, and S are 3, 2, 2, and 1. These letters appear with probabilities 0.15, 0.10, 0.40, and 0.35, where the letters underneath the bottle caps are independent of each other. Set up a simulation to estimate the expected value and the probability mass function of the number of bottles that must be purchased in order to form the word Babarras.

Chapter 6

A Gentle Introduction to Markov Chains

This chapter introduces you to the very basic concepts and features of Markov chains. A Markov chain is basically a sequence of random variables evolving over time and have a weak form of dependency between them. This very useful model was developed in 1906 by Russian mathematician A.A. Markov (1856–1922).[32] In a famous paper written in 1913, he used his probability model to analyze the frequencies at which vowels and consonants occur in Pushkin's novel "Eugene Onegin." Markov showed empirically that adjacent letters in Pushkin's novel are not independent but obey his theory of dependent random variables. The characteristic property of a Markov chain is that its memory goes back only to the most recent state. The Markov chain model is a very powerful probability model that is used today in countless applications in many different areas, such as voice recognition, DNA analysis, stock control, telecommunications, and a host of others.

6.1 Markov chain model

A Markov chain can be seen as a dynamic stochastic process that randomly moves from state to state with the property that only

---

[32] Markov lived through a period of great political activity in Russia and, having firm opinions as left wing activist, he became heavily involved. In 1913, the Romanov dynasty, which had been in power in Russia since 1613, celebrated their 300 years of power. Markov disapproved of the celebration and instead celebrated 200 years of the Law of Large Numbers!

the current state is relevant for the next state. In other words, the memory of the process goes back only to the most recent state. A picturesque illustration of this would show the image of a frog jumping from lily pad to lily pad with appropriate transition probabilities that depend only on the position of the last lily pad visited. In order to plug a specific problem into a Markov chain model, the state variable(s) should be appropriately chosen in order to ensure the characteristic memoryless property of the process. The basic steps of the modeling approach are:

- Choosing the state variable(s) such that the current state summarizes everything about the past that is relevant to the future states.

- The specification of the one-step transition probabilities of moving from state to state in a single step.

Using the concept of state and choosing the state in an appropriate way, surprisingly many probability problems can be solved within the framework of a Markov chain. In this chapter it is assumed that the set of states to be denoted by I is *finite*. This assumption is important. The theory of Markov chains involves quite some subtleties when the state space is countably infinite.

The following notation is used for the one-step transition probabilities:

$$p_{ij} = \text{the probability of going from state } i \text{ to state } j \text{ in one step}$$

for $i, j \in I$. Note that the one-step probabilities must satisfy $p_{ij} \geq 0$ for all $i, j \in I$ and $\sum_{j \in I} p_{ij} = 1$ for all $i \in I$.

Example 6.1. A faulty digital video conferencing system has a clustering error pattern. If a bit is received correctly, the probability of receiving the next bit correctly is 0.99. This probability is 0.95 if the last bit was received incorrectly. What is an appropriate Markov chain model?

Solution. The choice of the state is obvious in this example. Let state 0 mean that the last bit sent is not received correctly, and state

1 mean that the last bit sent is received correctly. The sequence of states is described by a Markov chain with one-step transition probabilities $p_{00} = 1-0.95$, $p_{01} = 0.95$, $p_{10} = 1-0.99$, and $p_{11} = 0.99$.

The choice of an appropriate state is more tricky in the next example. Putting yourself in the shoes of someone who must write a simulation program for the problem in question may be helpful in choosing the state variable(s).

Example 6.2. An absent-minded professor drives every morning from his home to the office and at the end of the day from the office to home. At any given time, his driver's license is located at his home or at the office. If his driver's license is at his location of departure, he takes it with him with probability 0.5. What is an appropriate Markov chain model for this situation?

Solution. Your first thought might be to define two states 1 and 0, where state 1 describes the situation that the professor has his driver's license with him when driving his car, and state 0 describes the situation that he does not have his driver's license with him when driving his car. However, these two states do not suffice for a Markov model: state 0 does not provide enough information to predict the state at the next drive (why?). In order to give the probability distribution of this next state, you need information about the current location of the driver's license of the professor. You get a Markov model by simply inserting this information into the state description. Therefore, the following three states are defined:

- state 1 = the professor is driving his car and has his driver's license with him,

- state 2 = the professor is driving his car and his driver's license is at the point of departure,

- state 3 = the professor is driving his car and his driver's license is at his destination.

This state description has the Markovian property that the present state contains sufficient information for predicting future states. The

Markov chain model has state space $I = \{1, 2, 3\}$. What are the one-step transition probabilities p_{ij}? The only possible one-step transitions from state 1 are to the states 1 and 2 (verify!). A one-step transition from state 1 to state 1 occurs if the professor does not forget his license on the next trip and so $p_{11} = 0.5$. By a similar argument, $p_{12} = 0.5$. Obviously, $p_{13} = 0$. A one-step transition from state 2 is always to state 3, and so $p_{23} = 1$ and $p_{21} = p_{22} = 0$. The only possible transitions from state 3 are to the states 1 and 2, and so $p_{33} = 0$. A one-step transition from state 3 to state 1 occurs if the professor takes his license with him on the next trip, and so $p_{31} = 0.5$. Similarly, $p_{32} = 0.5$. A *matrix* is the most useful way to display the one-step transition probabilities:

$$
\begin{array}{c|ccc}
\text{from}\backslash\text{to} & 1 & 2 & 3 \\
\hline
1 & 0.5 & 0.5 & 0 \\
2 & 0 & 0 & 1 \\
3 & 0.5 & 0.5 & 0
\end{array}.
$$

Time-dependent analysis of Markov chains

In Markov chains, a key role is played by the n-step transition probabilities. For any $n = 1, 2, \ldots$, these probabilities are defined as

$p_{ij}^{(n)}$ = the probability of going from state i to state j in n steps

for all $i, j \in I$. Note that $p_{ij}^{(1)} = p_{ij}$. How to calculate the n-step transition probabilities? It will be seen that these probabilities can be calculated by matrix products. Many calculations for Markov chains can be boiled down to matrix calculations

The so-called Chapman-Kolmogorov equations for calculating the n-step transition probabilities $p_{ij}^{(n)}$ are

$$
p_{ij}^{(n)} = \sum_{k \in I} p_{ik}^{(n-1)} p_{kj} \quad \text{for all } i, j \in I \text{ and } n = 2, 3, \ldots.
$$

This recurrence relation says that the probability of going from state i to state j in n steps is obtained by summing the probabilities of the mutually exclusive events of going from state i to some state k

in the first $n - 1$ steps and then going from state k to state j in the nth step. A formal proof proceeds as follows. For initial state i, let A be the event that the state is j after n steps of the process and B_k be the event that the state is k after $n - 1$ steps. Then, by the law of conditional probability, $P(A) = \sum_{k \in I} P(A \mid B_k) P(B_k)$. Since $P(A) = p_{ij}^{(n)}$, $P(B_k) = p_{ik}^{(n-1)}$ and $P(A \mid B_k) = p_{kj}$, the Chapman-Kolmogorov equations follow.

Let's now verify that the n-step transition probabilities can be calculated by multiplying the matrix of one-step transition probabilities by itself n times. To do so, let $\mathbf{P}$ be the matrix with the p_{ij} as entries and $\mathbf{P}^{(m)}$ be the matrix with the $p_{ij}^{(m)}$ as entries. In matrix notation, the Chapman-Kolmogorov equations read as

$$\mathbf{P}^{(n)} = \mathbf{P}^{(n-1)} \times \mathbf{P} \quad \text{for } n = 2, 3, \ldots .$$

This gives $\mathbf{P}^{(2)} = \mathbf{P}^{(1)} \times \mathbf{P} = \mathbf{P} \times \mathbf{P}$. Next, you get $\mathbf{P}^{(3)} = \mathbf{P}^{(2)} \times \mathbf{P} = \mathbf{P} \times \mathbf{P} \times \mathbf{P}$. Continuing in this way, you see that $\mathbf{P}^{(n)}$ is given by the n-fold matrix product $\mathbf{P} \times \mathbf{P} \times \cdots \times \mathbf{P}$, shortly written as $\mathbf{P}^n$. Thus, $\mathbf{P}^{(n)} = \mathbf{P}^n$, which verifies that

$p_{ij}^{(n)}$ is the (i, j)th entry of the n-fold matrix product $\mathbf{P}^n$.

Example 6.3. On the Island of Hope the weather each day is classified as sunny, cloudy, or rainy. The next day's weather depends only on today's weather and not on the weather of the previous days. If the present day is sunny, the next day will be sunny, cloudy, or rainy with probabilities 0.70, 0.10, and 0.20. The transition probabilities for the weather are 0.50, 0.25, and 0.25 when the present day is cloudy, and they are 0.40, 0.30, and 0.30 when the present day is rainy. What is the probability that it will be sunny three days from now if it is cloudy today? What is the probability distribution of the weather on a given day far away?

Solution. These questions can be answered by using a three-state Markov chain. Let's say that the weather is in state S if it is sunny, in state C if it is cloudy, and in state R if it is rainy. The evolution of the weather is described by a Markov chain with state space $I =$

$\{S, C, R\}$. The matrix **P** of the one-step transition probabilities of this Markov chain is given by

$$
\begin{array}{cccc}
\text{from}\backslash\text{to} & S & C & R \\
S & \begin{pmatrix} 0.70 & 0.10 & 0.20 \\
C & 0.50 & 0.28 & 0.22 \\
R & 0.40 & 0.30 & 0.30 \end{pmatrix}
\end{array}.
$$

To find the probability of having sunny weather three days from now, you need the matrix product

$$
\mathbf{P}^3 = \begin{pmatrix} 0.6018 & 0.1728 & 0.2254 \\ 0.5928 & 0.1805 & 0.2266 \\ 0.5864 & 0.1857 & 0.2279 \end{pmatrix}.
$$

From this matrix you read off that the probability of having sunny weather three days from now is $p_{CS}^{(3)} = 0.5928$ if it is cloudy today. What is the probability distribution of the weather on a day far away? Intuitively, you expect that this probability distribution does not depend on the present state of the weather. This is indeed the case. Trying several values of n, it was found after $n = 7$ matrix multiplications that the elements of the matrix $\mathbf{P}^7$ agree row-to-row to four decimal places:

$$
\mathbf{P}^n = \begin{pmatrix} 0.5967 & 0.1771 & 0.2262 \\ 0.5967 & 0.1771 & 0.2262 \\ 0.5967 & 0.1771 & 0.2262 \end{pmatrix} \qquad \text{for all } n \geq 7.
$$

Thus, the weather on a day far away will be sunny, cloudy, or rainy with probabilities of about 59.7%, 17.7%, and 22.6%, regardless of the present weather. It is intuitively obvious that these probabilities also give the long-run proportions of time that the weather will be sunny, cloudy, or rainy, respectively.

Many probability problems, which are seemingly unrelated to Markov chains, can be modeled as a Markov chain with the help of a little imagination. This is illustrated with the next example. This example nicely shows that the line of thinking through the concepts of state, and state transition is very useful to analyze the problem (and many other problems in applied probability!).

Example 6.4. Six fair dice will be simultaneously rolled. What is the probability mass function of the number of different outcomes that will show up?

Solution. In solving a probability problem, it is often helpful to recognize when two problems are equivalent even if they sound different. To put the dice problem into the framework of a Markov chain, consider the equivalent problem of repeatedly rolling a single die six times (the reformulated problem is an instance of the coupon collector's problem). For the reformulated problem, define the state of the process as the number of different face values seen so far. The evolution of the state is described by a Markov chain with state space $I = \{0, 1, \ldots, 6\}$, where state 0 is the initial state. The one-step transition probabilities are given by (verify!):

$$p_{01} = 1, \; p_{ii} = \frac{i}{6} \text{ and } p_{i,i+1} = 1 - \frac{i}{6} \text{ for } 1 \le i \le 5,$$

$$p_{66} = 1, \text{ and the other } p_{ij} = 0.$$

The sought probability mass function is obtained by calculating the probability $p_{0k}^{(6)}$ for $k = 1, \ldots, 6$, which gives the probability of getting exactly k different face values when rolling a single die six times. Multiplying the matrix $\mathbf{P}$ of one-step transition probabilities six times by itself results in the matrix

$$\mathbf{P}^6 = \begin{pmatrix} 0 & 0.0001 & 0.0199 & 0.2315 & 0.5015 & 0.2315 & 0.0154 \\ 0 & 0.0000 & 0.0068 & 0.1290 & 0.4501 & 0.3601 & 0.0540 \\ 0 & 0 & 0.0014 & 0.0570 & 0.3475 & 0.4681 & 0.1260 \\ 0 & 0 & 0 & 0.0156 & 0.2165 & 0.5248 & 0.2431 \\ 0 & 0 & 0 & 0 & 0.0878 & 0.4942 & 0.4180 \\ 0 & 0 & 0 & 0 & 0 & 0.3349 & 0.6651 \\ 0 & 0 & 0 & 0 & 0 & 0 & 1 \end{pmatrix}.$$

The first row of $\mathbf{P}^{(6)}$ gives the sought probabilities $p_{01}^{(6)} = 0.0001$, $p_{02}^{(6)} = 0.0199$, $p_{03}^{(6)} = 0.2315$, $p_{04}^{(6)} = 0.5015$, $p_{05}^{(6)} = 0.2315$, $p_{06}^{(6)} = 0.0154$. The tail probability $1 - p_{06}^{(6)} = 0.9846$ gives the probability that more than 6 rolls of a die will be needed to get all six possible outcomes. In the same way, you can use a Markov chain to calculate the tail probabilities for the number of purchases needed to get a complete collection in the general coupon collector's problem with equally likely coupons.

Problem 6.1. Consider Example 6.2 again. It is Wednesday evening and the professor is driving to home, unaware of the fact that there will be traffic control on the roadway to his house coming Friday evening. What is the probability that the professor will be fined for not having his license with him given that he left his license at the university on Wednesday evening? (answer: 0.625)

Problem 6.2. Every day, it is either sunny or rainy on Rainbow Island. The weather for the next day depends only on today's weather and yesterday's weather. The probability that it will be sunny tomorrow is 0.9 if the last two days were sunny, is 0.45 if the last two days were rainy, is 0.7 if today's weather is sunny and yesterday's weather was rainy, and is 0.5 if today's weather is rainy and yesterday's weather was sunny. Define a Markov chain that describes the weather on Rainbow Island and specify the one-step transition probabilities. What is the probability of having sunny weather five days from now if it rained today and yesterday? (answer: 0.7440)

Problem 6.3. An airport bus deposits 10 passengers at 7 stops. Each passenger is as likely to get off at any stop as at any other, and the passengers act independently of one another. The bus makes a stop only if someone wants to get off. Use Markov chain analysis to calculate the probability mass function of the number of bus stops. (answer: $(0.0000, 0.0000, 0.0069, 0.1014, 0.3794, 0.4073, 0.1049)$)

Problem 6.4. Consider Example 6.3 again. Use indicator random variables to calculate the expected value of the number of sunny days in the coming seven days given that it is cloudy today. (answer: 4.049)

6.2 Absorbing Markov chains

A powerful trick in Markov chain analysis is to use one or more absorbing states. A state i of a Markov chain is said to be *absorbing* if

$p_{ii} = 1$, that is, once the process enters an absorbing state i, it always stays there. Absorbing Markov chains are very useful to analyze success runs. This is illustrated with the following two examples.

Example 6.3 (continued). What is probability there will be three or more consecutive days with sunny weather in the coming 14 days given that it is rainy today?

Solution. Augment the three states in the Markov chain model from Example 6.3 with two additional states SS and SSS, where state SS means that it was sunny the last two days, and state SSS means that it was sunny the last three days. State SSS is taken as an absorbing state. The matrix $\mathbf{P}$ of one-step transition probabilities now becomes

$$
\begin{array}{c}
\text{from}\backslash\text{to} \\
S \\
C \\
R \\
SS \\
SSS
\end{array}
\begin{array}{ccccc}
S & C & R & SS & SSS
\end{array}
\left(
\begin{array}{ccccc}
0 & 0.10 & 0.20 & 0.70 & 0 \\
0.50 & 0.28 & 0.22 & 0 & 0 \\
0.40 & 0.30 & 0.30 & 0 & 0 \\
0 & 0.10 & 0.20 & 0 & 0.70 \\
0 & 0 & 0 & 0 & 1
\end{array}
\right).
$$

Since the process stays in state SSS once it is there, three consecutive days with sunny weather occur somewhere in the coming 14 days if and only if the process is in state SSS 14 days hence. Thus, $p_{R,SSS}^{(14)}$ gives the probability that there will three or more consecutive days with sunny weather in the coming 14 days given that it is rainy today. The matrix $\mathbf{P}^{(14)}$ is obtained by multiplying the matrix $\mathbf{P}$ by itself 14 times and is given by

$$
\mathbf{P}^{(14)} = \left(
\begin{array}{ccccc}
0.0214 & 0.0186 & 0.0219 & 0.0178 & 0.9203 \\
0.0320 & 0.0278 & 0.0328 & 0.0265 & 0.8809 \\
0.0344 & 0.0290 & 0.0342 & 0.0277 & 0.8757 \\
0.0117 & 0.0102 & 0.0120 & 0.0097 & 0.9564 \\
0 & 0 & 0 & 0 & 1
\end{array}
\right).
$$

You read off from row 3 that the desired probability equals 0.8757.

Example 6.5. What is the probability of getting either five or more consecutive heads or five or more consecutive tails, or both, in 25 tosses of a fair coin?

Solution. Use a Markov chain with six states $0, 1, \ldots, 5$. State 0 corresponds to the start of the coin-tossing experiment. For $i = 1, \ldots, 5$, state i means that the last i tosses constitute a run of length i, where a run consists of only heads or only tails. State 5 is taken as an absorbing state. The one-step transition probabilities of the Markov chain are $p_{01} = 1$, $p_{i,i+1} = p_{i1} = 0.5$ for $1 \leq i \leq 4$, $p_{55} = 1$, and the other $p_{ij} = 0$. The probability of getting either five or more consecutive heads or five or more consecutive tails, or both, in 25 tosses is $p_{05}^{(25)}$. Calculating

$$\mathbf{P}^{25} = \begin{pmatrix} 0 & 0.2336 & 0.1212 & 0.0629 & 0.0326 & 0.5496 \\ 0 & 0.2252 & 0.1168 & 0.0606 & 0.0314 & 0.5659 \\ 0 & 0.2089 & 0.1084 & 0.0562 & 0.0292 & 0.5974 \\ 0 & 0.1774 & 0.0920 & 0.0478 & 0.0248 & 0.6580 \\ 0 & 0.1168 & 0.0606 & 0.0314 & 0.0163 & 0.7748 \\ 0 & 0 & 0 & 0 & 0 & 1 \end{pmatrix},$$

you find that $p_{05}^{(25)} = 0.5496$. A remarkably high probability! Most people grossly underestimate the lengths of longest runs.

Example 6.6. Joe Dalton desperately wants to raise his bankroll of \$600 to \$1 000 in order to pay his debts before midnight. He enters a casino to play European roulette. He decides to bet on red each time using bold play, that is, Joe bets either his entire bankroll or the amount needed to reach the target bankroll, whichever is smaller. Thus the stake is \$200 if his bankroll is \$200 or \$800, and the stake is \$400 if his bankroll is \$400 or \$600. In European roulette a bet on red is won with probability $\frac{18}{37}$ and is lost with probability $\frac{19}{37}$. What is the probability that Joe will reach his goal?

Solution. Two approaches will be given to calculate the probability of Joe reaching his goal. The problem is modeled by a Markov chain with the six states $i = 0, 1, \ldots, 5$, where state i means that Joe's bankroll is $i \times 200$ dollars. The states 0 and 5 are absorbing and the game is over as soon one of these states is reached. Thus $p_{00} = p_{55} = 1$. The other p_{ij} are easily found. For example, the only possible one-step transitions from state $i = 2$ are to the states 0 and 4 because

Joe bets \$400 in state 2. Thus, $p_{20} = \frac{19}{37}$ and $p_{24} = \frac{18}{37}$. The other p_{ij} are in the matrix $\mathbf{P}$ of one-step transition probabilities:

$$
\begin{array}{c}
\text{from}\backslash\text{to} \\
0 \\
1 \\
2 \\
3 \\
4 \\
5
\end{array}
\begin{array}{cccccc}
0 & 1 & 2 & 3 & 4 & 5
\end{array}
\left(
\begin{array}{cccccc}
1 & 0 & 0 & 0 & 0 & 0 \\
\frac{19}{37} & 0 & \frac{18}{37} & 0 & 0 & 0 \\
\frac{19}{37} & 0 & 0 & 0 & \frac{18}{37} & 0 \\
0 & \frac{19}{37} & 0 & 0 & 0 & \frac{18}{37} \\
0 & 0 & 0 & \frac{19}{37} & 0 & \frac{18}{37} \\
0 & 0 & 0 & 0 & 0 & 1
\end{array}
\right).
$$

For any starting state, the process will ultimately be absorbed in either state 0 or state 5. The absorption probabilities can be obtained by calculating $\mathbf{P}^n$ for n sufficiently large.[33] Trying several values of n, it was found that $n = 20$ is large enough to have convergence of all $p_{ij}^{(n)}$ in four or more decimals. In particular, $p_{15}^{(20)} = 0.1859$, $p_{25}^{(20)} = 0.3820$, $p_{35}^{(20)} = 0.5819$ and $p_{45}^{(20)} = 0.7853$. Thus, the probability of Joe reaching his goal when starting with \$600 is 0.5819. This probability is the maximum probability of Joe for reaching his goal. The intuitive explanation that bold play is optimal in Joe's situation is that the shorter Joe exposes his bankroll to the casino's house advantage, the better it is (e.g., if Joe bets \$50 each time, he reaches his goal with probability 0.4687).

Alternatively, the probability of Joe reaching his goal can be obtained by solving four linear equations. To do so, define f_i as the probability of getting absorbed in state 5 when the starting state is i. By definition, $f_0 = 0$ and $f_5 = 1$. By conditioning on the next state after state i, you get the linear equations

$$
f_1 = \frac{19}{37} \times f_0 + \frac{18}{37} \times f_2, \quad f_2 = \frac{19}{37} \times f_0 + \frac{18}{37} \times f_4,
$$
$$
f_3 = \frac{19}{37} \times f_1 + \frac{18}{37} \times f_5, \quad f_4 = \frac{19}{37} \times f_3 + \frac{18}{37} \times f_5,
$$

giving $f_1 = 0.1859$, $f_2 = 0.3820$, $f_3 = 0.5819$, and $f_4 = 0.7853$.

---

[33]The computational effort to calculate $\mathbf{P}^n$ for large n can be considerably reduced by the trick $\mathbf{P}^4 = \mathbf{P}^2 \times \mathbf{P}^2$, $\mathbf{P}^8 = \mathbf{P}^4 \times \mathbf{P}^4$, $\mathbf{P}^{16} = \mathbf{P}^8 \times \mathbf{P}^8$, and so on.

Problem 6.5. What is the probability of getting five or more consecutive heads during 25 tosses of a fair coin? (answer: 0.3116)

Problem 6.6. Consider Problem 5.25 again. For $n = 5$, 10, 15, and 25, calculate the probability that more than n trials are needed until all dice are banked. (answer: 0.7860, 0.4102, 0.1824, and 0.0311)

Problem 6.7. What is the probability of getting two consecutive totals of 7 before a total of 12 when repeatedly rolling two dice? (answer: 0.4615)

Problem 6.8. You toss a fair coin until HTH or HHT appears. What is the probability that HHT appears first? (answer: $\frac{2}{3}$)

Problem 6.9. For the Venice-53 lottery in Section 5.3, formulate an absorbing Markov chain to calculate the probability that it takes more than n draws before number 53 shows up.

Problem 6.10. You have three whiskey glasses labeled 1, 2, and 3. Initially, all glasses are filled. A fair die is rolled. If the outcome is 1 or 6, then glass 1 is emptied if it is full and is filled if it is empty. The same happens to glass 2 if a 2 or a 5 is rolled, and to glass 3 if a 3 or a 4 is rolled. What is the probability that more than 10 dice rolls are needed until all three glasses are empty? (answer: 0.3660)

Problem 6.11. In the dice game of Pig, you repeatedly roll a single die. Upon rolling a 1, your turn is over, and you get a score zero. Otherwise, you can stop whenever you want and then your score is the total number of points rolled. Under the hold-at-20 rule, you stop when you have rolled 20 points or more.[34] Use Markov chain analysis to get the probability mass function of your end score under the hold-at-20 rule. (answer: (0.6245, 0.0997, 0.0950, 0.0742, 0.0542, 0.0352, 0.0172) on (0, 20, 21, 22, 23, 24, 25) with E(end score) = 8.14)

---

[34]The rationale behind this stopping rule is as follows. Suppose your current score is x points and you decide for one other roll of the die. Then, the expected value of the change of your score is $\sum_{k=2}^{6} \frac{1}{6} \times k - \frac{1}{6} \times x = \frac{20}{6} - \frac{x}{6}$, which is non-positive for $x \geq 20$. This is the principle of the one-stage-look-ahead rule.

Problem 6.12. You are fighting a dragon with two heads. Each time you swing at the dragon with your sword, there is a 75% chance of knocking off one head and a 50% chance of missing. If you miss, either one additional head or two additional heads will grow immediately before you can swing again at the dragon. The probability of one additional head is 0.7 and of two additional heads is 0.3. You win if you have knocked off all of the dragon's heads, but you must run for your life if the dragon has five or more heads. Use Markov chain analysis to calculate your chance of winning. (answer: 0.5210)

Problem 6.13. Consider Problem 5.27 from Chapter 5 again. Use Markov chain analysis to compute the probability that the guest will lose the game. (answer: $\frac{5}{9}$)

6.3 The gambler's ruin problem

A nice illustration of an absorbing Markov chain is the gambler's ruin problem that goes back to Christiaan Huygens (1629–1695) and Blaise Pascal (1623–1662). This random walk problem will be used to demonstrate that the absorption probabilities can also be calculated by solving a system of linear equations instead of taking matrix products. The method of linear equations can also be used to calculate the expected time until absorption. The gambler's ruin problem is as follows. Two players A and B with initial bankrolls of a dollars and b dollars play a game until one of the players is bankrupt. In a play of the game, player a wins one dollar from player B with probability p and loses one dollar to player B with probability $q = 1 - p$. The successive plays of the game are independent of each other. What is the probability $P(a, b)$ that player A is the ultimate winner, and what is the expected value $E(a, b)$ of the number of plays until one of the players goes broke?

The quantities $P(a, b)$ and $E(a, b)$ can be found by using an absorbing Markov chain. The Markov chain has the states $0, 1, \ldots, a+b$, where state i means that the current bankroll of player A is i dollars (and that of player B is $a + b - i$ dollars). The states 0 and $a + b$ are taken as absorbing states. The other one-step transition probabilities are given by $p_{i,i+1} = p$, $p_{i,i-1} = q$, and $p_{ij} = 0$ otherwise for

$i = 1, 2, \ldots, a + b - 1$. The probability $P(a, b)$ that player A will be the ultimate winner can be found by calculating the n-step transition probability $p_{a,a+b}^{(n)}$ for sufficiently large values of n. However, a more elegant approach is as follows. For $i = 0, 1, \ldots, a + b$, define f_i as the probability that the Markov chain will be ultimately absorbed in state $a + b$ when the starting state is i. By definition, $f_0 = 0$ and $f_{a+b} = 1$. The other f_i can be found by solving the linear equations

$$f_i = pf_{i+1} + qf_{i-1} \quad \text{for } i = 1, 2, \ldots, a + b - 1.$$

In particular, f_a gives the desired probability $P(a, b)$. The equation for f_i is easily explained from the law of conditional probability: the term pf_{i+1} accounts for the case of a win of player A in state i and the term qf_{i-1} for a win of player B in state i. Since the equations for the f_i are so-called linear difference equations, they can be explicitly solved. The details are omitted and we state only the famous *gambler's ruin formula* for $f_a = P(a, b)$:

$$P(a, b) = \begin{cases} \dfrac{1 - (q/p)^a}{1 - (q/p)^{a+b}} & \text{if } p \neq q \\ \dfrac{a}{a+b} & \text{if } p = q. \end{cases}$$

To get $E(a, b)$, define e_i as the expected value of the number of transitions of the Markov chain needed to reach either state 0 or state $a + b$ when starting from state i. By definition, $e_0 = e_{a+b} = 0$. The other e_i can be found by solving the linear equations

$$e_i = 1 + pe_{i+1} + qe_{i-1} \quad \text{for } i = 1, 2, \ldots, a + b - 1.$$

To derive these equations, let Z_i be the remaining number of plays when i is the current state. Under the condition that player A wins the next play, the conditional distribution of Z_i is the same as the distribution of $1 + Z_{i+1}$; otherwise, it is the same as the distribution of $1 + Z_{i-1}$. Thus, by the law of conditional expectation, $E(Z_i) = pE(1 + Z_{i+1}) + qE(1 + Z_{i-1})$. The resulting equations for the $e_i = E(Z_i)$ can be explicitly solved. In particular, by $E(a, b) = e_a$,

$$E(a, b) = \begin{cases} \dfrac{a}{q-p} - \dfrac{a+b}{q-p} \dfrac{1 - (q/p)^a}{1 - (q/p)^{a+b}} & \text{if } p \neq q \\ ab & \text{if } p = q. \end{cases}$$

To illustrate the gambler's ruin formula, suppose you go to the casino of Monte Carlo with 100 euro, and your goal is to double it. You opt to play European roulette, betting each time on red. You will double your stake with probability $p = \frac{18}{37}$, and you will lose with probability $q = \frac{19}{37}$. So, if you stake \$5 each time $(a = b = 20)$, or \$10 $(a = b = 10)$, or \$25 $(a = b = 4)$, or \$50 $(a = b = 2)$, the probability of reaching your goal will have the values of 0.2533, 0.3680, 0.4461, and 0.4730, respectively. The expected number of bets has the values 365.19, 97.66, 15.94, and 4.00, respectively.

6.4 Long-run behavior of Markov chains

Let's consider a Markov chain with a finite set of states I (the assumption of a finite state space is important in the steady-state analysis of Markov chains). The process starts in one of the states and moves successively from one state to another, where the probability of moving from state i to state j in one step is denoted by p_{ij}. What about the probability distribution of the state after many, many transitions? Does the effect of the starting state ultimately fade away? In order to answer these questions, two conditions are introduced.

Condition C_1. The Markov chain has no two or more disjoint closed sets of states, where a set C is said to be closed if $p_{ij} = 0$ for $i \in C$ and $j \notin C$.

Condition C_2. The set of states cannot be split into multiple disjoint sets $S_1, \ldots, S_d$ with $d \geq 2$ such that a one-step transition from a state in S_k is always to a state in S_{k+1}, where $S_{d+1} = S_1$.

The condition C_1 is satisfied in nearly any application, but this is not the case for condition C_2. The condition C_2 rules out periodicity in the state transitions such as is the case in the three-state Markov chain with $p_{12} = p_{13} = 0.5$, $p_{21} = p_{31} = 1$, and $p_{ij} = 0$ otherwise. In this Markov chain, $p_{11}^{(n)}$ is alternately 0 or 1 and so $\lim_{n \to \infty} p_{11}^{(n)}$ does not exist (neither any of the $p_{ij}^{(n)}$ has a limit as n tends to infinity).

Under the conditions C_1 and C_2 the equilibrium probability)

$$\pi_j = \lim_{n \to \infty} p_{ij}^{(n)}$$

exists for all $i, j \in I$ and is independent of the starting state i. The π_j can be calculated as the unique solution to the balance equations

$$\pi_j = \sum_{k \in I} p_{kj} \pi_k \quad \text{for } j \in I$$

together with the *normalization equation* $\sum_{j \in I} \pi_j = 1$. The balance equations can be easily explained from the Chapman-Kolmogorov equations $p_{ij}^{(n)} = \sum_{k \in I} p_{ik}^{(n-1)} p_{kj}$. Letting $n \to \infty$ in both sides of the Chapman-Kolmogorov equations and interchanging limit and summation (justified by the finiteness of I), you get the balance equations. The reader is asked to take for granted that the limits π_j exist and are uniquely determined by the set of linear equations.

As an illustration, consider Example 6.3 again. The balance equations are then given by

$$\pi_S = 0.70\pi_S + 0.50\pi_C + 0.40\pi_R,$$
$$\pi_C = 0.10\pi_S + 0.28\pi_C + 0.30\pi_R,$$
$$\pi_R = 0.20\pi_S + 0.22\pi_C + 0.30\pi_R.$$

In solving these equations together with $\pi_S + \pi_C + \pi_R = 1$, you are allowed to delete one of the balance equations in order to get a square system of linear equations. Solving the equations gives $\pi_S = 0.5967$, $\pi_C = 0.1771$, and $\pi_R = 0.2262$. The same numerical answers as were obtained in Example 6.3 by calculating $\mathbf{P}^n$ for large n.

Interpretation of the equilibrium probabilities

The equilibrium probability π_j can be interpreted as the probability that you will find the Markov chain in state j many, many transitions later, whatever the current state is. Essential to this interpretation is that you are not given any information about the course of the process between now and the distant future. A second interpretation of π_j is that of a long run average: the proportion of transitions to state j will be π_j when averaging over many, many transitions. This second interpretation remains valid when the non-periodicity Condition C_2 is not satisfied. In general, it is true that

$$\pi_j = \lim_{n \to \infty} \frac{1}{n} \sum_{k=1}^{n} p_{ij}^{(k)}$$

exists for all $i, j \in I$ and is independent of the starting state i (this so-called Césaro limit is equal to the ordinary limit if the latter exists). Again, the π_j are the unique solution to the balance equations $\pi_j = \sum_{k \in I} p_{kj} \pi_k$ for $j \in I$ together with the equation $\sum_{j \in I} \pi_j = 1$.

The following example deals with a periodic Markov chain and is prelude to material on reversibility in Section 6.5.

Example 6.7. Two compartments A and B together contain r particles. With the passage of every time unit, one of the particles is selected at random and is removed from its compartment to the other. What is the equilibrium distribution of the number of particles in compartment A?

Solution. The process describing the number of particles in compartment A is a Markov chain with state space $I = \{0, 1, \ldots r\}$. The one-step transition probabilities are $p_{i,i+1} = \frac{r-i}{r}$ for $i = 0, 1, \ldots, r-1$, $p_{i,i-1} = \frac{i}{r}$ for $i = 1, 2, \ldots, r$, and $p_{ij} = 0$ otherwise. This Markov model is known as the Ehrenfest model in physics. The equilibrium equations are

$$\pi_j = \pi_{j-1} \frac{r-j+1}{r} + \pi_{j+1} \frac{j+1}{r} \quad \text{for } j = 1, \ldots, r-1,$$

with $\pi_0 = \frac{1}{r}\pi_1$ and $\pi_r = \frac{1}{r}\pi_{r-1}$. Intuitively, any marked particle is to be found equally likely in either of the two compartments after many transitions. This suggests that $\{\pi_j\}$ is a binomial distribution. Indeed, by substitution, it is readily verified that $\pi_j = \binom{r}{j}(\frac{1}{2})^r$ for $j = 0, 1, \ldots, r$. The Markov chain is periodic: a transition from any state in the subset of even-numbered states leads to a state in the subset of odd-numbered states, and vice versa. Hence, the probability π_j can only be interpreted as the long-run proportion of time during which compartment A contains j particles. It is interesting to note that the average number of transitions from state i to state $i-1$ per unit time must be equal to the average number of transitions from state $i-1$ to state i per unit time in the long run for any i (between any two transitions from i to $i-1$ there must be a transition from $i-1$ to i, and vice versa, as a consequence of $p_{ij} = 0$ for $|i - j| > 1$). The Ehrenfest model is an example of a so-called reversible Markov chain, which concept will be discussed in Section 6.5.

Page-ranking algorithm

The equilibrium theory of Markov chains has many applications. The most famous application is Markov's own analysis of the frequencies at which vowels and consonants occur in Pushkin's novel "Eugene Onegin," see also Problem 6.14 below.

An important application of more recent date is the application of Markov chains to the ranking of web pages. The page-ranking algorithm is one of the methods Google uses to determine a page's relevance or importance. Suppose that you have n interlinked web pages. Let n_j be the number of outgoing links on page j. It is assumed that $n_j > 0$ for all j. Let α be a given number with $0 < \alpha < 1$. Imagine that a random surfer jumps from his current page by choosing with probability α a random page amongst those that are linked from the current page, and by choosing with probability $1 - \alpha$ a completely random page. Hence the random surfer jumps around the web from page to page according to a Markov chain with the one-step transition probabilities

$$p_{jk} = \alpha\, r_{jk} + (1 - \alpha)\, \frac{1}{n} \quad \text{for } j, k = 1, \ldots, n,$$

where $r_{jk} = \frac{1}{n_j}$ if page k is linked from page j and $r_{jk} = 0$ otherwise. The parameter α was originally set to 0.85. The inclusion of the term $(1 - \alpha)/n$ can be justified by assuming that the random surfer occasionally gets bored and then randomly jumps to any page on the web. Since the probability of such a jump is rather small, it is reasonable that it does not influence the ranking very much. By the term $(1 - \alpha)/n$ in the p_{jk}, the Markov chain has no two or more disjoint closed sets and is aperiodic. Thus, the Markov chain has a unique equilibrium distribution $\{\pi_j\}$. These probabilities can be estimated by multiplying the matrix $\mathbf{P}$ of one-step transition probabilities by itself repeatedly. Because of the constant $(1 - \alpha)/n$ in the matrix $\mathbf{P}$, things mix better up so that the n-fold matrix product $\mathbf{P}^n$ converges very quickly to its limit. The equilibrium probability π_j gives us the long-run proportion of time that the random surfer will spend on page j. If $\pi_j > \pi_k$, then page j is more important than page k and should be ranked higher.

Problem 6.14. In a famous paper written in 1913, Andrey Markov analyzed an unbroken sequence of 20 000 letters from the poem Eugene Onegin. He found that the probability of a vowel following a vowel is 0.128, and that the probability of a vowel following a consonant is 0.663. Use a Markov chain to estimate the percentages of vowels and consonants in the novel. (answer: 43.2% and 56.8%)

Problem 6.15. What is for Example 6.1 the long-run fraction of bits that are incorrectly received? (answer: $\frac{1}{96}$)

Problem 6.16. In a small college town, there are an Italian, a Mexican, and a Thai restaurant. A student eating at the Italian restaurant will eat the following evening at the Italian restaurant with probability 0.10, in the Mexican restaurant with probability 0.35, at the Thai restaurant with probability 0.25, or at home with probability 0.30. The probabilities of switching are 0.4, 0.15, 0.25, and 0.2 when eating at the Mexican restaurant, 0.5, 0.15, 0.05, and 0.3 when eating at the Thai restaurant, and 0.40, 0.35, 0.25, and 0 when eating at home. What proportion of time the student will eat at home? (answer: 0.2110)

Problem 6.17. In a certain town, there are four entertainment venues. Both Linda and Bob are visiting every weekend one of these venues, independently of each other. Each of them visits the venue of the week before with probability 0.4 and chooses otherwise at random one of the other three venues. What is the long-run fraction of weekends that Linda and Bob visit a same venue? (answer: $\frac{1}{4}$)

Problem 6.18. Consider the following modification of Example 6.2. In case the driver's license of the professor is at his point of departure, the professor takes it with him with probability 0.75 when departing from home and with probability 0.5 when departing from the office. What is the long-run fraction of time the professor has his license with him? (answer: 0.4286)

Problem 6.19. Consider Problem 6.2 again. What is the long-run fraction of sunny days? (answer: 0.7912) The entrepreneur Jerry Wood has a pub on the island. On every sunny day, his sales are

$N(\mu_1, \sigma_1^2)$ distributed with $\mu_1 = \$1\,000$ and $\sigma_1 = \$200$, while on rainy days his sales are $N(\mu_2, \sigma_2^2)$ distributed with $\mu_2 = \$500$ and $\sigma_2 = \$75$. What is the long-run average sales per day? (answer: $\$895.60$)

6.5 Markov chain Monte Carlo simulation[35]

This section gives a first introduction to Markov chain Monte Carlo (MCMC). This method can be used to tackle computational problems that arise among others in Bayesian inference. Let S be a very large but finite set on which a probability mass function $\pi(s)$ is given that is only known up to a multiplicative constant. It is not feasible to compute the constant directly. How to calculate $\sum_{s \in S} h(s)\pi(s)$ for a given function $h(s)$? The idea is to construct a Markov chain that has $\pi(s)$ as its equilibrium distribution and to simulate a sequence $s_1, s_2, \ldots, s_m$ of successive states of this Markov chain for large m. Then $\sum_{s \in S} h(s)\pi(s)$ can be estimated by $\frac{1}{m}\sum_{k=1}^{m} h(s_k)$.

Reversible Markov chains

The concept of reversible Markov chains plays a crucial role in Markov chain Monte Carlo simulation. Let's consider an *irreducible* Markov chain with finite state space I and one-step transition probabilities p_{ij}. Irreducibility means that each state can be reached from any other state, that is, for any two states i and j there is a $n \geq 1$ such that $p_{ij}^{(n)} > 0$. Since an irreducible Markov chain satisfies Condition C_1, it has a unique equilibrium distribution $\{\pi_j\}$. The irreducible Markov chain is said to be *reversible* if the equilibrium probabilities π_j satisfy the so-called *detailed balance equations*

$$\pi_j p_{jk} = \pi_k p_{kj} \quad \text{for all } j, k \in I.$$

These equations have the following physical interpretation: the average number of transitions from state j to state k per unit time is equal to the average number of transitions from state k to state j per unit time in the long-run for any $j, k \in I$. The following result

---

[35]This section contains advanced material. This material is taken from Henk Tijms, *Probability, a Lively Introduction*, Cambridge University Press, 2017.

will be crucial. In the Metropolis–Hastings algorithm below, the following result is crucial. If a probability distribution $\{a_j\}$ on I can be constructed such that

$$a_j p_{jk} = a_k p_{kj} \quad \text{for all } j, k \in I,$$

then $\{a_j\}$ gives the equilibrium distribution $\{\pi_j\}$ of the Markov chain. This is easy to prove. Sum both sides of the equation $a_j p_{jk} = a_k p_{kj}$ over $k \in I$. Together with $\sum_{k \in I} p_{jk} = 1$, this gives

$$a_j = \sum_{k \in I} a_k p_{kj} \quad \text{for all } j \in I.$$

These equations are precisely the equilibrium equations of the Markov chain and so, by the uniqueness of the equilibrium distribution, $a_j = \pi_j$ for all j.

6.5.1 Metropolis–Hastings algorithm

The Metropolis–Hastings algorithm is an example of a Markov chain Monte Carlo method. The algorithm will be first explained for the case of a discrete probability distribution, but the basic idea of the algorithm can be directly generalized to the case of a continuous probability distribution.

Let S be a very large but finite set of states on which a probability mass function $\{\pi(s),\ s \in S\}$ is given, where $\pi(s) > 0$ for all s, and the $\pi(s)$ are only known up to a multiplicative constant. The Metropolis–Hastings algorithm generates a sequence of states $(s_1, s_2, \ldots)$ from a Markov chain that has $\{\pi(s),\ s \in S\}$ as its unique equilibrium distribution. To that end, the algorithm uses a candidate-transition function $q(t \mid s)$ (for clarity of presentation, the notation $q(t \mid s)$ is used rather than p_{st}). This function is to be interpreted as saying that when the current state is s, the candidate for the next state is t with probability $q(t \mid s)$. Thus, you first choose, for each $s \in S$, a probability mass function $\{q(t \mid s),\ t \in S\}$. These functions must be chosen in such a way that the Markov matrix with the $q(t \mid s)$ as one-step transition probabilities is irreducible. The idea is to next adjust these transition probabilities in such a way that the resulting Markov chain has $\{\pi(s),\ s \in S\}$ as unique equilibrium distribution. The reversibility equations are the key to this idea. If the candidate-transition function $q(t \mid s)$ already satisfies the detailed balance equations

$$\pi(s)q(t \mid s) = \pi(t)q(s \mid t) \quad \text{for all } s, t \in S,$$

you are done: the Markov chain with the $q(t \mid s)$ as one-step transition probabilities is reversible and has $\{\pi(s)\}$ as its unique equilibrium distribution. What should you do when the detailed balance equations are not fully satisfied? The answer is to modify the one-step transition probabilities by rejecting certain transitions. To work out this idea, fix two states s and t for which the detailed balance equation is not satisfied. There is no restriction to assume that $\pi(s)q(t \mid s) > \pi(t)q(s \mid t)$. Otherwise, reverse the roles of the states s and t. If $\pi(s)q(t \mid s) > \pi(t)q(s \mid t)$, then, loosely speaking, the process moves from s to t too often. How could you restore this? A simple trick to reduce the number of transitions from s to t is to use an acceptance probability $\alpha(t \mid s)$: the process is allowed to make the transition from s to t with probability $\alpha(t \mid s)$ and otherwise the process stays in the current state s. The question remains how to choose $\alpha(t \mid s)$. The choice of $\alpha(t \mid s)$ is determined by the requirement

$$\pi(s)[q(t \mid s)\alpha(t \mid s)] = \pi(t)[q(s \mid t)\alpha(s \mid t)].$$

Taking $\alpha(s \mid t) = 1$ for transitions from t to s, you get

$$\alpha(t \mid s) = \frac{\pi(t)q(s \mid t)}{\pi(s)q(t \mid s)}.$$

Therefore, for any $s, t \in S$, the acceptance probability is defined by

$$\boxed{\alpha(t \mid s) = \min\left[\frac{\pi(t)q(s \mid t)}{\pi(s)q(t \mid s)}, 1\right].}$$

Finally, the one-step transition probabilities to be used in the algorithm are defined by

$$\boxed{q_{MH}(t \mid s) = \begin{cases} q(t \mid s)\alpha(t \mid s) & \text{for } t \neq s \\ 1 - \sum_{t \neq s} q(t \mid s)\alpha(t \mid s) & \text{for } t = s. \end{cases}}$$

The Markov chain with these one-step transition probabilities satisfies the detailed balance equations $\pi(s)q_{MH}(t \mid s) = \pi(t)q_{MH}(s \mid t)$ for all s, t. Therefore, this Markov chain has $\{\pi(s), s \in S\}$ as its unique equilibrium distribution. It is important to note that for the construction of the Markov chain, it suffices to know the π_s up to a multiplicative constant because the acceptance probabilities involve only the ratio's $\pi(s)/\pi(t)$.

Summarizing, the Markov chain operates as follows. If the current state is s, a candidate state t is generated from the probability mass function $\{q(t \mid s), t \in S\}$. If $t \neq s$, then state t is accepted with probability $\alpha(t \mid s)$ as the next state of the Markov chain; otherwise, the Markov chain stays in state s.

Metropolis–Hastings algorithm

Step 0. Choose probability mass functions $\{q(t \mid s), t \in S\}$ for $s \in S$ such that the Markov matrix with the $q(t \mid s)$ as elements is irreducible. Choose a starting state s_0. Let $n := 1$.
Step 1. Generate a candidate state t_n from the probability mass function $\{q(t \mid s_{n-1}), t \in S\}$. Calculate the acceptance probability

$$\alpha = \min \left[\frac{\pi(t_n)q(s_{n-1} \mid t_n)}{\pi(s_{n-1})q(t_n \mid s_{n-1})}, 1 \right].$$

Step 2. Generate a random number u from $(0, 1)$. If $u \leq \alpha$, accept t_n and let $s_n := t_n$; otherwise, $s_n := s_{n-1}$.
Step 3. $n := n + 1$. Repeat Step 1 with s_{n-1} replaced by s_n.

Note that when the chosen probability densities $q(t \mid s)$ are symmetric, that is, $q(t \mid s) = q(s \mid t)$ for all $s, t \in S$, then the acceptance probability α in Step 1 reduces to

$$\alpha = \min \left(\frac{\pi(t_n)}{\pi(s_{n-1})}, 1 \right).$$

In applications of the algorithm, one typically want to estimate $E[h(X)]$ for a given function $h(x)$, where X is a random variable with $\pi(s)$ as its probability density. If the states $s_1, s_2, \ldots, s_m$ are generated by the Metropolis–Hastings algorithm for a sufficiently large m, then $E[h(X)] = \sum_{s \in S} h(s)\pi(s)$ is estimated by

$$\frac{1}{m} \sum_{k=1}^{m} h(s_k).$$

This estimate is based on a law of large numbers result for Markov chains, saying that $\lim_{m \to \infty} \frac{1}{m} \sum_{k=1}^{m} h(s_k) = E[h(X)]$. A heuristic explanation is as follows. The probability $\pi(s)$ can be interpreted as the long-run fraction of transitions into state s and so, for large m, $\pi(s) \approx \frac{m(s)}{m}$, where $m(s)$ is the number of times that state s occurs among the sequence $s_1, s_2, \ldots, s_m$. This gives $\sum_{s \in S} h(s)\pi(s) \approx \frac{1}{m} \sum_{s \in S} h(s)m(s) = \frac{1}{m} \sum_{k=1}^{m} h(s_k)$.

The Metropolis–Hastings algorithm directly extends to the case of a probability density $\pi(s)$ on a (multi-dimensional) continuous set S, where you want to calculate $\int_{s \in S} h(s)\pi(s)\, ds$ for the case that the density $\pi(s)$ is only known up to a multiplicative constant.

What are the best options for the proposal functions $q(t \mid s)$? There are two general approaches:

(a) In *independent chain sampling* the candidate state t is drawn independently of the current state s of the Markov chain, that is, $q(t \mid s) = g(t)$ for some proposal density $g(x)$. It is important that the tail of the proposal density $g(s)$ dominates the tail of the target density $\pi(s)$ in order to have a well-mixing Markov chain.

(b) In *random walk chain sampling* the candidate state t is the current state s plus a draw from a random variable Z that does not depend on the current state. In this case, $q(t \mid s) = g(t - s)$ with $g(z)$ the density of the random variable Z.

6.5.2 Gibbs sampler

The Gibbs sampler is another Markov chain Monte Carlo method. It is frequently used in Bayesian statistics. Gibbs sampling is applicable when a multivariate probability density function is not known explicitly or is difficult to sample from, whereas the univariate conditional probability densities are known and are easy to sample from. For ease of presentation, consider a three-component joint probability density $\pi(x, y, z)$ of a random vector (X, Y, Z) whose univariate conditional probability densities $\pi_X(x \mid y, z)$, $\pi_Y(y \mid x, z)$, and $\pi_Z(z \mid x, y)$ are fully known.[36] Starting with arbitrarily chosen values (x_0, y_0, z_0), the Gibbs sampler iteratively updates one of the components of the state (x, y, z). There are two common schemes to determine which component to update. One scheme is to choose the component randomly, the other known as the standard Gibbs sampler is to choose the component by sequentially scanning through the components.

Standard Gibbs sampling goes as follows. If the current state is (x_n, y_n, z_n) after the nth iteration, the state is updated at the $(n + 1)$th iteration according to the scheme

    a. Sample x_{n+1} from $\pi_X(x \mid y_n, z_n)$.
    b. Sample y_{n+1} from $\pi_Y(y \mid x_{n+1}, z_n)$.
    c. Sample z_{n+1} from $\pi_Z(z \mid x_{n+1}, y_{n+1})$.

At each step of the sampling, the most recent values of the other components are used in the univariate conditional densities. Letting $\pi^{(n)}(x, y, z)$ and $\pi_X^{(n)}(x)$, $\pi_Y^{(n)}(b)$, $\pi_Z^{(n)}(z)$ be the estimates of $\pi(x, y, z)$ and the marginal univariate densities $\pi_X(x)$, $\pi_Y(y)$, $\pi_Z(z)$, it can be shown under conditions of irreducibility and aperiodicity that these estimates converge to $\pi(x, y, z)$ and $\pi_X(x)$, $\pi_Y(y)$, $\pi_Z(z)$ as $n \to \infty$.

Example 6.8. In an actuarial model, the vector (X, Y, Z) has a trivariate probability density $\pi(x, y, z)$ that is proportional to

$$\binom{z}{x} y^{x+\alpha-1}(1 - y)^{z-x+\beta-1} e^{-\lambda} \frac{\lambda^z}{z!}$$

for $x = 0, 1, \ldots, z$, $0 < y < 1$, and $z = 0, 1, \ldots$. The random variable Z is the number of policies in a portfolio, the random variable Y is the claim probability for any policy, and the random variable X is the number of policies resulting in a claim. Take the data $\alpha = 2$, $\beta = 8$, and $\lambda = 50$. How to estimate the expected value, the standard deviation, and the marginal density of the random variable X?

---

[36]For clarification, take the case of discrete random variables. Then $\pi_X(x \mid y, z)$ is defined as $P(X = x \mid Y = y$ and $Z = z)$ and can be written as the ratio of $P(X = x$ and $Y = y$ and $Z = z)$ and $\sum_s P(X = s$ and $Y = y$ and $Z = z)$.

Solution. The estimates can be found by the Gibbs sampler. The univariate conditional densities can be explicitly determined by using

$$\pi_X(x \mid y, z) = \frac{\pi(x, y, z)}{\sum_{s=0}^{z} \pi(s, y, z)}, \quad \pi_Y(y \mid x, z) = \frac{\pi(x, y, z)}{\int_{s=0}^{1} \pi(x, s, z)\,ds},$$

$$\text{and } \pi_Z(z \mid x, y) = \frac{\pi(x, y, z)}{\sum_{s=x}^{\infty} \pi(x, y, s)}.$$

A little algebra shows that $\pi_X(x \mid y, z)$ is proportional to $y^x(1-y)^{z-x}$ as function of x with $x = 0, 1, \ldots, z$, $\pi_Y(y \mid x, z)$ is proportional to $y^{x+\alpha-1}(1-y)^{z-x+\beta-1}$ as function of y with $0 < y < 1$, and $\pi_Z(z \mid x, y)$ is proportional to $[\lambda(1-y)]^{z-x}/(z-x)!$ as function of z with $z = x, x+1, \ldots$. Thus, $\pi_X(x \mid y, z)$ is the binomial density with parameters z and y, $\pi_Y(y \mid x, z)$ is the beta distribution with parameters $x+\alpha$ and $z-x+\beta$, and $\pi_Z(z \mid x, y)$ is the Poisson density shifted to x and having parameter $\lambda(1-y)$. Next, it is straightforward to apply the Gibbs sampler. Ready-to-use codes to simulate from specific probability distributions are available in languages as Python and R. The estimates 9.992 and 6.836 are found for $E(X)$ and $\sigma(X)$ after 100 000 iterations (the exact values are 10 and 6.809). Figure 18 gives the simulated histogram for the marginal density of X.

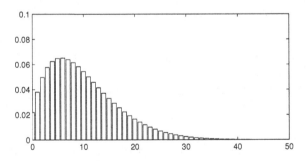

Figure 18: Gibbs results for the density of X.

Solutions to Selected Problems

Fully worked-out solutions to a number of problems are given. Including worked-out solutions is helpful for students who use the book for self-study and stimulates active learning. Make sure you try the problems before looking to the solutions.

2.4. A useful trick is to imagine that both the balls and the boxes are labeled as 1, 2, and 3. The sample space consists of $3^3 = 27$ equally likely outcomes (b_1, b_2, b_3), where b_i is the label of the box into which the ball with label i falls. The number of outcomes in which only a specified box remains empty is $2 \times \binom{3}{2} = 6$: two possibilities for the box that gets two balls and $\binom{3}{2}$ possibilities for these two balls. Thus, the total number of outcomes in which exactly one box remains empty is $3 \times 6 = 18$. Thus, the sought probability is $\frac{18}{27} = \frac{2}{3}$.

2.5. It is often helpful to rephrase a probability problem in another context. The probability that the other card has no free drinks either is nothing else than the probability of getting 10 heads and 10 tails in 20 coin tosses (think about it!). To calculate the latter probability, use as sample space the set consisting of all possible sequences of H's and T's of length 20. The sample space has 2^{20} equally likely elements. There are $\binom{20}{10}$ sequences having 10 H's and 10 T's. Thus, the probability of getting 10 heads and 10 tails is $\binom{20}{10}/2^{20} = 0.1762$. Therefore, the sought-after probability is 0.1762.

2.6. The strategy is that the daughter (father) opens door 1 (2) first. If the key (car) is behind door 1 (2), the daughter (father) goes on to open door 2 (1). If the goat is after the first opened door, door 3 is opened

215

as second. Then four of the six possible configurations of the car, key, and goat are favorable: the configurations (car, key, goat), (car, goat, key), (key, car, goat), and (goat, key, car) are winning, and the two configurations (key, goat, car) and (goat, car, key) are losing.

2.7. If the sets A and B are not disjoint, then $P(A) + P(B)$ counts twice the probability of the set of outcomes that belong to both A and B. Therefore, $P(A \text{ and } B)$ should be subtracted from $P(A) + P(B)$. Let A be the event that the chosen card is a heart and B be the event that it is an ace, then $P(A) = \frac{13}{52}$, $P(B) = \frac{4}{52}$ and $P(A \text{ and } B) = \frac{1}{52}$, and so $P(A \text{ or } B) = \frac{13}{52} + \frac{4}{52} - \frac{1}{52} = \frac{16}{52}$.

Note: The formula for $P(A \text{ or } B)$ can be directly extended to

$$P(A \text{ or } B \text{ or } C) = P(A) + P(B) + P(C) - P(A \text{ and } B)$$
$$- P(A \text{ and } C) - P(B \text{ and } C) + P(A \text{ and } B \text{ and } C).$$

More generally,

$$P(A_1 \text{ or } A_2 \text{ or } \ldots \text{ or } A_r) = \sum_{i=1}^{r} P(A_i) - \sum_{i<j} P(A_i \text{ and } A_j)$$

$$+ \sum_{i<j<k} P(A_i \text{ and } A_j \text{ and } A_k) - \ldots + (-1)^{r+1} P(A_1 \text{ and } \ldots \text{ and } A_r).$$

This is the *inclusion-exclusion formula*. This formula is very useful in combinatorial probability. As an illustration, suppose that balls are put in three bins numbered as $i = 1$, 2, and 3 until each bin contains one or more balls. Each ball is put in bin i with probability p_i, where $p_1 = 0.2$, $p_2 = 0.3$, and $p_3 = 0.5$. What is the probability Q_n that more than n balls are needed to get at least one ball in each bin? For fixed n, let A_i be the event that bin i is still empty after putting n balls in the bins. Then $Q_n = P(A_1 \text{ or } A_2 \text{ or } A_3)$. Since $P(A_i) = (1 - p_i)^n$, $P(A_i \text{ and } A_j) = (1-p_i-p_j)^n$ for $i \neq j$ and $P(A_1 \text{ and } A_2 \text{ and } A_3) = 0$, the inclusion-exclusion formula gives

$$Q_n = \sum_{i=1}^{3} (1 - p_i)^n - \sum_{i=1}^{2} \sum_{j=i+1}^{3} (1 - p_i - p_j)^n \qquad \text{for all } n \geq 3.$$

As sanity check, take $n = 3$. The alternative calculation $Q_3 = 1 - 3! p_1 p_2 p_3$ gives the same answer 0.81 as the inclusion-exclusion formula.

2.8. The probability is $\sum_{k=1}^{\infty} (1 - p_1 - p_2)^{k-1} p_1 = \frac{p_1}{p_1 + p_2} = 0.40$.

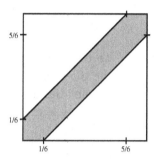

Figure 19: Meeting problem.

2.9. Translate the problem into choosing a point at random inside the unit square. The probability that the two persons will meet within 10 minutes of each other is equal to the probability that a point chosen at random in the unit square will fall inside the shaded region, see Figure 19. The area of the shaded region is calculated as $1 - \frac{5}{6} \times \frac{5}{6} = \frac{11}{36}$. Dividing this by the area of the unit square, you get that the desired probability is $\frac{11}{36}$.

2.11. Imagine that the two cards are picked one by one. Then, using the product rule, the probability of getting two red cards is $\frac{14}{21} \times \frac{13}{20} = \frac{13}{30}$, and the probability of getting one red card and one black card is $\frac{14}{21} \times \frac{7}{20} + \frac{7}{21} \times \frac{14}{20} = \frac{14}{30}$. Alternatively, the probability of getting two red cards is $\binom{14}{2} / \binom{21}{2} = \frac{13}{30}$, and the probability of getting one red card and one black card is $\binom{14}{1}\binom{7}{1} / \binom{21}{2} = \frac{14}{30}$.

2.14. For the case of 4 boiled eggs and 2 raw eggs, the probability that the person who begins smashes a raw egg as first is $\frac{2}{6} + \frac{4}{6} \times \frac{3}{5} \times \frac{2}{4} + \frac{4}{6} \times \frac{3}{5} \times \frac{2}{4} \times \frac{1}{3} \times 1 = 0.6$. For the other case, the probability is $\frac{1}{6} + \frac{1}{6} + \frac{1}{6} = \frac{1}{2}$.

2.22. Let A be the event that Arthur wins the match and B_i be the event that Arthur loses i games of the first two for $i = 0, 1$, and 2. By the law of conditional probability,

$$P(A) = P(A \mid B_0)P(B_0) + P(A \mid B_1)P(B_1) + P(A \mid B_2)P(B_2)$$
$$= 1 \times p^2 + P(A) \times 2pq + 0 \times q^2.$$

This gives $P(A) = \frac{p^2}{1-2pq} = \frac{p^2}{p^2+q^2}$, where the last equality uses the fact that $p^2 + 2pq + q^2 = (p+q)^2 = 1$.

2.28. Let the hypothesis H be the event that the standard die was picked and let the evidence E_1 be the event that the first roll of the die has the outcome 6. The prior probabilities are $P(H) = P(\overline{H}) = \frac{1}{2}$, and the likelihood ratio has $P(E_1 \mid H) = \frac{1}{6}$ and $P(E_1 \mid \overline{H}) = \frac{1}{3}$. Thus, the posterior odds of the hypothesis H are

$$\frac{P(H \mid E_1)}{P(\overline{H} \mid E_1)} = \frac{1/2}{1/2} \times \frac{1/6}{1/3} = \frac{1}{2},$$

which gives the updated value $\frac{1/2}{1+1/2} = \frac{1}{3}$ for the probability that the standard die was picked. The second question can be answered in two ways. After the first roll has been done but before the second roll will be done, you take the posterior probabilities $P(H \mid E) = \frac{1}{3}$ and $P(\overline{H} \mid E) = \frac{2}{3}$ as the prior probabilities for $P(H)$ and $P(\overline{H})$. Doing so and letting E_2 be the event that the second roll has outcome 6, you get

$$\frac{P(H \mid E_2)}{P(\overline{H} \mid E_2)} = \frac{1/3}{2/3} \times \frac{1/6}{1/3} = \frac{1}{4},$$

and so the newly updated value of the probability that the standard die was picked is $\frac{1}{5}$. Alternatively, this probability can be calculated by letting the evidence $E_{1,2}$ be the event that each of the first two rolls of the picked die has outcome 6. Then, before the first two rolls are done, the priors $P(H)$ and $P(\overline{H})$ are $\frac{1}{2}$, and the likelihood ratio has $P(E_{1,2} \mid H) = \frac{1}{6} \times \frac{1}{6}$ and $P(E_{1,2} \mid \overline{H}) = \frac{1}{3} \times \frac{1}{3}$. This leads to

$$\frac{P(H \mid E_{1,2})}{P(\overline{H} \mid E_{1,2})} = \frac{1/2}{1/2} \times \frac{1/36}{1/9} = \frac{1}{4},$$

which gives again the update $\frac{1}{5}$ for the probability that the standard die was picked. The Bayesian approach has the feature that you can continuously update your beliefs as information accrues. Verify yourselves that the updated value of the probability that the standard die was picked becomes $\frac{1}{9}$ after a third roll with outcome 2. *Note:* this problem nicely illustrates the Bayesian view that probabilities represent the knowledge an observer has about the state of nature of a physical object.

2.29. Let the hypothesis H be the event that the person has the disease and the evidence E be the event that he has tested positive. The prior probabilities are $P(H) = 0.001$ and $P(\overline{H}) = 0.999$. Also, the likelihood ratio has $P(E \mid H) = 0.99$ and $P(E \mid \overline{H}) = 0.01$. Thus, the posterior odds of hypothesis H are $\frac{0.001}{0.999} \times \frac{0.99}{0.01} = \frac{11}{111}$, and so the posterior probability of hypothesis H is $\frac{11/111}{1+11/111} = 0.0902$. In other words, the probability that the person has the disease is only 9.02% if the first test is positive. The low value of this probability may be surprising when not taking into account the *base rate*: most positive tests come from people who don't have the disease. If a second independent test also gives a positive test result, then use the posterior probabilities 0.0902 and $1 - 0.0902 = 0.9098$ as new prior probabilities for $P(H)$ and $P(\overline{H})$. This leads to the new posterior odds $\frac{0.0902}{0.9098} \times \frac{0.99}{0.01} = 9.815$. Thus, after a second positive test, the probability that the person has the disease is $\frac{9.815}{1+9.815} \times 100 = 90.75\%$.

The posterior probabilities can also be heuristically argued be the expected frequency approach. This is first done for the case that 0.1% of the test population has the disease. Think of a large number of people, say 10 000. Of these 10 000 people, 10 will have the disease on average, and 9 990 will not. Of the 10 persons with the disease $0.99 \times 10 = 9.9$ will test positive on average, and of the 9 990 persons with no disease $0.01 \times 9 990 = 99.9$ will test positive on average. Thus, the posterior probability of the disease is equal to $9.9/(9.9 + 99.9) = 0.0902$ when the first test is positive. Similarly, if a second test is also positive, you get the posterior probability $892.98/(892.98+90.98) = 0.9075$.

2.38. Let X be the number of cards that need to be turned over. Using conditional probabilities, $E(X) = \sum_{k=1}^{49} k a_k$, where $a_1 = \frac{4}{52}$ and

$$a_k = \left(1 - \frac{4}{52}\right) \times \cdots \times \left(1 - \frac{4}{52 - k + 2}\right) \times \frac{4}{52 - k + 1} \quad \text{for } k \geq 2.$$

You can write a_k as $\left[\binom{48}{k-1}/\binom{52}{k-1}\right] \times \frac{4}{52-k+1}$. This gives $E(X) = 10.6$. An alternative calculation is as follows. In a random shuffle, four aces divide the deck into five parts, each of which can contain from 0 to 48 non-ace cards. By a symmetry argument, all five parts must have the same expected length of $\frac{48}{5} = 9.6$ cards. Thus, the expected number of cards to be turned over until an ace appears for the ith time is $i \times (9.6 + 1) = i \times 10.6$ for $i = 1, \ldots, 4$.

2.39. $E(X) = m\binom{10}{m}/\binom{12}{m}$ is maximal for $m = 4$ with $E(X) = \frac{56}{33}$.

2.43. This problem is an application of the balls-and-bins model with $n = \binom{42}{6}$ bins (six-number combinations) and $b = 5\,000\,000$ balls (tickets). By the same reasoning as in Example 2.13, you find that the expected value of the number of empty bins is $n \times \left(\frac{n-1}{n}\right)^b$, which equals $2\,022\,388$ for $n = 5\,245\,786$ bins and $b = 5\,000\,000$ balls. Thus, the expected number of *different* six-number combinations filled in is $5\,245\,786 - 2\,022\,388 = 3\,22\,398$.

2.44. Using the algebraic identities $\sum_{k=1}^{n} k = \frac{1}{2}n(n+1)$ and $\sum_{k=1}^{n} k^2 = \frac{1}{6}n(n + 1)(2n + 1)$, the results are obtained after some algebra.

2.46. The formulas $\sum_{k=1}^{\infty} kx^{k-1} = \frac{1}{(1-x)^2}$ and $\sum_{k=2}^{\infty} k(k - 1)x^{k-2} = \frac{2}{(1-x)^3}$ for $|x| < 1$ are given in Section 1.2. The first formula shows that $E(X) = \sum_{k=1}^{\infty} k(1 - p)^{k-1}p = \frac{p}{(1-(1-p))^2} = \frac{1}{p}$. The second formula shows that $E[X(X - 1)] = \sum_{k=2}^{\infty} k(k - 1)(1 - p)^{k-1}p$ is equal to $\frac{2p(1-p)}{(1-(1-p))^3} = \frac{2(1-p)}{p^2}$. Thus, $E(X^2)$ equals $\frac{2(1-p)}{p^2} + \frac{1}{p} = \frac{2-p}{p^2}$, and so $\sigma^2(X) = \frac{2-p}{p^2} - \left(\frac{1}{p}\right)^2 = \frac{1-p}{p^2}$. Note: the geometric distribution is memoryless, that is, for any r, $P(X > r + k \mid X > r) = P(X > k)$ for $k = 1, 2, \ldots$, as is easily verified by using the fact that $P(X > j) = (1 - p)^j$ for all j.

2.47. Writing $\sum_{k=0}^{\infty} P(N > k)$ as $\sum_{k=0}^{\infty} \sum_{j=k+1}^{\infty} P(N = j)$, you get that $\sum_{k=0}^{\infty} P(N > k)$ equals $\sum_{j=1}^{\infty} \sum_{k=0}^{j-1} P(N = j) = \sum_{j=1}^{\infty} j\,P(N = j) = E(N)$, by an interchange of the order of summation. In the same way, you get the alternative formula for $E[N(N - 1)]$, using the identity $\sum_{k=0}^{j-1} k = \frac{1}{2}j(j - 1)$.

2.48. Let S be the number of purchases needed to get a complete set of cards. Then S can be written as $S = Y_0 + Y_1 + \cdots + Y_{49}$, where Y_i is the number of purchases needed to go from i distinct cards to $i+1$ distinct cards. Then Y_i is geometrically distributed with parameter $p_i = \frac{50-i}{50}$ and so $E(Y_i) = \frac{50}{50-i}$. This gives $E(S) = 50 \sum_{k=1}^{50} \frac{1}{k} = 224.96$. For the case of c equally likely cards, $E(S) = c \sum_{k=1}^{c} \frac{1}{k}$.

2.50. Suppose the strategy is to stop as soon as you have picked a number larger than or equal to r. The number of trials needed is geometrically distributed with success probability $p = \frac{25-r+1}{25}$ (and expected value $\frac{25}{25-r+1}$). Each of the values $r, r+1, \ldots, 25$ is equally likely for your payout. Thus, the expected net payoff is $\sum_{k=r}^{25} k \times \frac{1}{25-r+1} - \frac{25}{25-r+1} \times 1$, which can be simplified as $\frac{1}{2}(25+r) - \frac{25}{25-r+1}$. This expression has the maximal value 18.4286 for $r = 19$.

2.51. Let the number a be your guess and Y be the randomly chosen number. Your expected winning is $E[g(Y)]$, where $g(y) = a^2$ for $y > a$ and $g(y) = 0$ otherwise. Then $E[g(Y)] = \sum_{k=a+1}^{100} a^2 P(Y = k)$. Since $P(Y = k) = 1/100$ for all k, you get $E[g(Y)] = (100-a)a^2/100$. This expression is maximal for $a = 67$ with 1481.37 as maximum value.

2.54. Let the random variable S be number of purchases needed to get a complete set of cards. Then $S = \sum_{i=0}^{49} Y_i$, where Y_i is geometrically distributed with parameter $\frac{50-i}{50}$ and the Y_i are independent. Using the formula $\sigma^2(S) = \sum_{i=0}^{49} \sigma^2(Y_i)$ and results of Problem 2.46, $\sigma^2(S) = \sum_{i=0}^{49} \left(\frac{50}{50-i}\right)^2 (1 - \frac{50-i}{50}) = 3837.872$ and so $\sigma(S) = 61.951$. For the case of c equally likely cards, $\sigma^2(S) = c^2 \sum_{k=1}^{c} \frac{1}{k^2} - c \sum_{k=1}^{c} \frac{1}{k}$.

2.60. Since $P(X \geq c) = P(tX \geq tc) = P(e^{tX} \geq e^{tc})$ for any $t \geq 0$, you get the generic Chernoff bound by applying Markov's inequality with $Y = e^{tX}$ and $a = e^{tc}$. Noting that $P(X \leq c) = P(tX \geq tc)$ for $t < 0$, a repetition of the above argument gives the generic bound $P(X \leq c) \leq E(e^{tX})/e^{tc}$ for any $t < 0$. The bounds depend on the moment generating function $E(e^{tX})$ and the chosen value for t. There are many Chernoff bounds as a result. Chernoff bounds are particularly useful to bound tail probabilities of a finite sum of independent random variables. For example, suppose that $X = X_1 + \cdots + X_n$, where $X_1, \ldots, X_n$ are independent Bernoulli variables with $P(X_i = 1) = p_i$. Then, the following Chernoff bounds can be derived

$$P(X \geq a\mu) \leq e^{-g(a)\mu} \text{ for } a \geq 1, \ P(X \leq a\mu) \leq e^{-g(a)\mu} \text{ for } 0 < a < 1,$$

where $\mu = E(X) = \sum_{i=1}^{n} p_i$ and $g(a) = a \ln(a) - a + 1$. For $0 < \delta < 1$, these bounds are the basis for the simpler but weaker Chernoff bounds

$$P(X \geq (1+\delta)\mu) \leq e^{-\frac{1}{3}\delta^2\mu} \text{ and } P(X \leq (1-\delta)\mu) \leq e^{-\frac{1}{2}\delta^2\mu}.$$

The Chernoff bounds $P(X \geq \mu + a) \leq e^{-\frac{1}{2}a^2/n}$ and $P(X \leq \mu - a) \leq e^{-\frac{1}{2}a^2/n}$ for any $a > 0$ can be derived with $X = \sum_{k=1}^{n} X_k$ and $\mu = E(X)$, where $X_1, \ldots, X_n$ are independent random variables on $[0, 1]$. In computer science, Chernoff bounds are often used in the performance analysis of randomized algorithms.

2.62. Put for abbreviation $\mu_X = E(X)$, $\mu_Y = E(Y)$, $\sigma_X = \sqrt{\text{var}(X)}$, $\sigma_Y = \sqrt{\text{var}(Y)}$, and $\rho = \rho(X, Y)$. Writing $Y - (\alpha + \beta X)$ in the form $Y - \mu_Y - \beta(X - \mu_X) + (\mu_Y - \alpha - \beta\mu_X)$ and using the linearity of expectation, you get after some algebra that

$$E[(Y - (\alpha + \beta X))^2] = \sigma_Y^2 + \beta^2 \sigma_X^2 - 2\beta\text{cov}(X, Y) + (\mu_Y - \alpha - \beta\mu_X)^2.$$

Putting the partial derivatives with respect to α and β equal to zero and noting that $\text{cov}(X, Y) = \rho\sigma_X\sigma_Y$, you find that $\beta = \frac{\rho\sigma_Y}{\sigma_X}$ and $\alpha = \mu_Y - \frac{\rho\sigma_Y}{\sigma_X}\mu_X$, which gives the regression line $y = \mu_Y + \rho\frac{\sigma_Y}{\sigma_X}(x - \mu_X)$. *Note:* if n independent observations (x_j, y_j) are given, then, for n large, μ_X can be estimated by $\overline{x} = \frac{1}{n}\sum_{j=1}^{n} x_j$, μ_Y by $\overline{y} = \frac{1}{n}\sum_{j=1}^{n} y_j$, σ_X by $\overline{\sigma}_X = \frac{1}{n}\sum_{j=1}^{n}(x_j - \overline{x})^2$, σ_Y by $\overline{\sigma}_Y = \frac{1}{n}\sum_{j=1}^{n}(y_j - \overline{y})^2$, and $\rho(X, Y)$ by $\frac{1}{n}\sum_{j=1}^{n}(x_j - \overline{x})(y_j - \overline{y})/(\overline{\sigma}_X\overline{\sigma}_Y)$. The least squares regression line is trustworthy if the residuals — the differences between the observed values y_j and the fitted values $\hat{y}_j$ of the dependent variable Y — do not show a pattern but are randomly scattered around zero. The residual plot, plotting the residuals $y_j - \hat{y}_j$ against the x_j, provides a useful visual assessment of the appropriateness of the linear regression model.

2.63. The underlying sample space for (X, Y) is the set of the 36 equally likely sample points (i, j) for $i, j = 1, \ldots, 6$, where i is the score of the first die and j is the score of the second die. The variable Y is equal to 10 for the three sample points $(4,6)$, $(6,4)$, and $(5,5)$. Thus, $P(X = 5 \mid Y = 10) = \frac{1/36}{3/36} = \frac{1}{3}$ and $P(X = 6 \mid Y = 10) = \frac{2/36}{3/36} = \frac{2}{3}$, and so $E(X \mid Y = 10) = 5 \times \frac{1}{3} + 6 \times \frac{2}{3} = \frac{17}{3}$.

2.64. Since $P(X = x$ and $Y = y) = (y - x - 1)/\binom{100}{3}$ for $1 \leq x \leq 98$ and $x + 2 \leq y \leq 100$, you get the marginal distribution $P(Y = y) = \frac{1}{2}(y-1)(y-2)/\binom{100}{3}$ for $y = 3, \ldots, 100$ and the conditional distribution $P(X = x \mid Y = y) = \frac{2(y-x-1)}{(y-1)(y-2)}$. This leads to $E(X \mid Y = y) = \frac{2}{(y-1)(y-2)} \sum_{x=1}^{y-2} x(y - x - 1) = \frac{1}{3}y$.

3.7. Let the indicator variable I_k be 1 if the outcome k appears two or more times when rolling six dice and let I_k be 0 otherwise. Then $E(I_k) = P(I_k = 1)$, where $P(I_k = 1)$ is equal to $\sum_{j=2}^{6} \binom{6}{j}(\frac{1}{6})^j(\frac{5}{6})^{6-j} = 0.2632$. This gives $E\left(\sum_{k=1}^{6} I_k\right) = \sum_{k=1}^{6} E(I_k) = 1.579$.

3.9. Your bankroll after 10 bets is $1.7^k \times 0.5^{10-k} \times 100$ dollars if k of the 10 tosses result in heads. The largest value of k such that $1.7^k \times 0.5^{10-k} < 0.5$ is $k = 5$. The binomial probability of getting no more than 5 heads in 10 tosses is 0.6230. The lesson is: do not simply rely on averages in situations of risk, but use probability distributions!

3.10. For fixed n, let A be the event that some person has to pay for the beer when n friends are still in the game. Let $p_n = P(A)$. Then, $p_1 = 1$. By the law of conditional probability, $p_n = \sum_{k=0}^{n} P(A \mid B_k)P(B_k)$ for $n \geq 2$, where B_k is the event that k tails appear when n coins are tossed. Since $P(A \mid B_k) = p_{n-k}$ and $P(B_k) = \binom{n}{k}(\frac{1}{2})^n$, you get the recursion $p_n = \sum_{k=0}^{n-1} \binom{n}{k}(\frac{1}{2})^n p_{n-k}$ for $n = 2, 3, \ldots$. Starting with $p_1 = 1$, the p_n can be recursively computed. This gives $p_7 = 0.7211$. Recursive thinking can be very rewarding!

3.15. A bit of imagination shows that this problem can be translated into the urn model with $R = 500$ red and $W = 999\,498$ white balls, where the red balls represent the 500 acquaintances of the first person. The sought probability is given by the probability that at least one red ball will be drawn when 500 balls are randomly taken out of the urn. This probability is given by the ratio of $1 - \binom{999\,498}{500}$ and $\binom{999\,998}{500}$ and equals 0.2214. The probability of 22% is surprisingly large. Events are often less "coincidental" than we may tend to think!

3.24. The probability $P(X_1 = x_1$ and $\ldots$ and $X_b = x_b)$ equals

$$e^{-\lambda} \frac{\lambda^{x_1 + \cdots + x_b}}{(x_1 + \cdots + x_b)!} \times \binom{x_1 + \cdots + x_b}{x_1} \cdots \binom{x_{b-1} + x_b}{x_{b-1}} p_1^{x_1} \cdots p_b^{x_b}$$

$$= e^{-\lambda p_1} \frac{(\lambda p_1)^{x_1}}{x_1!} \times \cdots \times e^{-\lambda p_b} \frac{(\lambda p_b)^{x_b}}{x_b!}.$$

This implies that the random variables $X_1, \ldots, X_b$ are independent, where X_j is Poisson distributed with parameter λp_j for $j = 1, \ldots, b$. By the independence of the separate Poisson distributions, the Poissonized balls-and-bins model is computationally easy to handle. For example, the probability that each bin will contain at least one ball is

$$(1 - e^{\lambda p_1}) \times \cdots \times (1 - e^{\lambda p_b}).$$

Similarly, the probability that each bin will contain two or more balls is $(1 - e^{\lambda p_1} - \lambda p_1 e^{-\lambda p_1}) \times \cdots \times (1 - e^{\lambda p_b} - \lambda p_b e^{-\lambda p_b})$. The Poissonized balls-and-bins model in which the number of balls to be put into the bins is randomized and has a Poisson distribution with expected value $\lambda = n$ can be used to get approximate results for the classical balls-and-bins model in which a fixed number of n balls are put into the bins, where n is large. The Poisson distribution with expected value $\lambda = n$ is nearly symmetric around n and has most of its mass concentrated near n when n is large, and so it is reasonable to use the Poissonized balls-and-bins model as an approximation to the classical balls-and-bins model, see also Section 4.7 and Problem 5.35 which deal with the coupon collector's problem. This problem is a manifestation of the balls-and-bins model.

3.26. The length X of a gestation period is $N(\mu, \sigma^2)$ distributed with $\mu = 280$ days and $\sigma = 10$ days. The probability that a birth is more than 15 days overdue is $1 - P(X \leq 295)$. This probability can be evaluated as $1 - P\left(\frac{X-280}{10} \leq \frac{295-280}{10}\right) = 1 - \Phi(1.5) = 0.0668$.

3.29. Let X denote the demand for the item. The normally distributed random variable X has an expected value of $\mu = 100$ and satisfies $P(X > 125) = 0.05$. To find the unknown standard deviation σ of X, write $P(X > 125) = 0.05$ as $P\left(\frac{X-100}{\sigma} > \frac{125-100}{\sigma}\right) = 1 - \Phi\left(\frac{25}{\sigma}\right) = 0.05$. Thus, $\Phi\left(\frac{25}{\sigma}\right) = 0.95$. The percentile $\xi_{0.95} = 1.645$ is the unique solution to the equation $\Phi(x) = 0.95$. Thus, $\frac{25}{\sigma} = 1.645$, which gives $\sigma = 15.2$.

3.34. Letting $X_1, \ldots, X_n$ be independent random variables with $P(X_i = 1) = P(X_i = -1) = 0.5$ for all i, the random variable D_n can be represented as $D_n = |X_1 + \cdots + X_n|$. Since $E(X_i) = 0$ and $\sigma(X_i) = 1$ for all i, the expected value and standard deviation of $X_1 + \cdots + X_n$ are 0 and $\sqrt{n}$. By the central limit theorem, the random variable $X_1 + \cdots + X_n$ is approximately $N(0, n)$ distributed for large n. Let's now calculate $E(|V|)$ if V is $N(0, \sigma^2)$ distributed:

$$E(|V|) = \frac{1}{\sigma\sqrt{2\pi}} \int_{-\infty}^{\infty} |v| e^{-\frac{1}{2}v^2/\sigma^2} \, dv = \frac{2}{\sigma\sqrt{2\pi}} \int_0^{\infty} v e^{-\frac{1}{2}v^2/\sigma^2} \, dv.$$

You get $E(|V|) = \frac{\sigma}{\sqrt{2\pi}} \int_0^{\infty} e^{-\frac{1}{2}w} \, dw = \frac{2\sigma}{\sqrt{2\pi}}$, by the change of variable $w = (v/\sigma)^2$. This gives the desired result for $E(D_n)$.

3.36. Let the random variable X_i be the dollar amount the casino loses on the ith bet. The X_i are independent random variables with $P(X_i = 10) = \frac{18}{37}$ and $P(X_i = -5) = \frac{19}{37}$. Then $E(X_i) = \frac{85}{37}$ and $\sigma(X_i) = \frac{45}{37}\sqrt{38}$. The total dollar amount lost by the casino is $\sum_{i=1}^{2\,500} X_i$. By the central limit theorem, this sum is approximately $N(\mu, \sigma^2)$ distributed with $\mu = 2\,500 \times \frac{85}{37}$ and $\sigma = 50 \times \frac{45}{37}\sqrt{38}$. The casino will lose no more than 6 500 dollars with a probability of about $\Phi\left(\frac{6\,500-\mu}{\sigma}\right) = 0.978$.

3.38. By the memoryless property of the exponential distribution, your waiting time at the bus stop is more than s minutes with probability $e^{-\frac{1}{10}s}$. Solving $e^{-\frac{1}{10}s} = 0.05$ gives $s = 29.96$. Thus, you should leave home about 7:10 a.m.

3.39. The result follows by noting that $P(t < X \leq t + \Delta t \mid X > t)$ equals $(e^{-\lambda t} - e^{-\lambda(t+\Delta)t})/e^{-\lambda t} = 1 - e^{-\lambda \Delta t} = \lambda \Delta t + o(\Delta t)$, where $o(\Delta t)$ is the generic symbol for a term that is negligibly small compared to Δt as Δt tends to zero.

Note: An alternative definition of a Poisson process with rate λ is: for any $t > 0$, the probability of exactly one event in a very small time interval $(t, t + \Delta t)$ is $\lambda \Delta t + o(\Delta t)$, the probability of zero events in the interval is $1 - \lambda \Delta t + o(\Delta t)$, and the probability of two or more events in the interval is $o(\Delta t)$ itself, independent of the history of the process before time t. The alternative definition of the Poisson process can be extended to the situation in which the occurrence of events is time-dependent. A counting process $\{N(t), t \geq 0\}$ is called a *non-homogeneous Poisson process* with event rate function $\lambda(t)$ if

(a) the numbers of events in disjoint time intervals are independent.
(b) $P(N(t + \Delta t) - N(t) = 1) = \lambda(t)\Delta t + o(\Delta t)$ as $\Delta t \to 0$.
(c) $P(N(t + \Delta t) - N(t) \geq 2) = o(\Delta t)$ as $\Delta t \to 0$.

If the event rate function $\lambda(t)$ is bounded in t, the non-homogeneous Poisson process can be constructed from a homogeneous Poisson process with rate λ, where $\lambda \geq \lambda(t)$ for all t. If an event occurring at time s is accepted with probability $\frac{\lambda(t)}{\lambda}$ and is rejected otherwise, then the counting process $\{N(t), t \geq 0\}$ with $N(t)$ the number of accepted events up to time t is a non-homogeneous Poisson process with event rate function $\lambda(t)$. The explanation is simple: the probability of an accepted event in $(t, t + \Delta t)$ is

$$\lambda\Delta t \times \frac{\lambda(t)}{\lambda} + o(\Delta t) = \lambda(t)\Delta t + o(\Delta t) \text{ as } \Delta t \to 0.$$

This construction enables you to simulate a non-homogeneous Poisson process.

3.42. Your win probability is the probability of having exactly one signal in (s, T). This probability is $e^{-\lambda(T-s)}\lambda(T - s)$. Putting the derivative of $e^{-\lambda(T-s)}\lambda(T - s)$ with respect to s equal to zero, you get that the optimal value of s is $T - \frac{1}{\lambda}$. The maximal win probability is e^{-1}.

3.43. Imagine that there is an infinite queue of cars waiting to be served. Then, service completions occur according to Poisson process with a rate of $1/15$ car per minute. The sought probability is the probability of no more than one service completion in 20 minutes, and is thus equal to the Poisson probability $e^{-20/15} + e^{-20/15}\frac{20/15}{1!} = 0.6151$.

3.47. Let the random variable X_i be the score on exam i for $i = 1, 2$.
(a) The density of X_1 is the $N(\mu_1, \sigma_1^2) = N(75, 12)$ density. This gives

$$P(X_1 \geq 80) = 1 - \Phi\left(\frac{80 - 75}{12}\right) = 0.3385.$$

(b) The random variable $X_1 + X_2$ is $N(\mu_1 + \mu_2, \sigma_1^2 + \sigma_2^2 + 2\rho\sigma_1\sigma_2) = N(140, 621)$ distributed. Thus,

$$P(X_1 + X_2 > 150) = 1 - \Phi\left(\frac{150 - 140}{\sqrt{621}}\right) = 0.3441.$$

(c) The random variable $X_2 - X_1$ is $N(\mu_2 - \mu_1, \sigma_1^2 + \sigma_2^2 - 2\rho\sigma_1\sigma_2) = N(-10, 117)$ distributed. Thus,

$$P(X_2 > X_1) = 1 - \Phi\left(\frac{10}{\sqrt{117}}\right) = 0.1776.$$

(d) The conditional distribution of X_2 given that $X_1 = 80$ equals the $N(65 + 0.7 \times \frac{15}{12} \times (80 - 75), (1 - 0.49) \times 225) = N(69.375, 114.75)$ distribution. Thus,

$$P(X_2 > X_1 \mid X_1 = 80) = 1 - \Phi\left(\frac{80 - 69.375}{\sqrt{114.75}}\right) = 0.1606.$$

6.1. $p_{22}^{(4)} + p_{23}^{(4)} = 0.625.$

6.2. Use a Markov model with four states SS, SR, RS, and RR describing the weather of yesterday and today. The one-step transition probabilities are

from\to	SS	SR	RS	RR
SS	0.9	0.1	0	0
SR	0	0	0.5	0.5
RS	0.7	0.3	0	0
RR	0	0	0.45	0.55

The probability of having sunny weather five days from now if it rained both today and yesterday is $p_{RR,SS}^{(5)} + p_{RR,RS}^{(5)}$. Calculating

$$\mathbf{P}^5 = \begin{pmatrix} 0.7178 & 0.0977 & 0.0879 & 0.0966 \\ 0.6151 & 0.1024 & 0.1325 & 0.1501 \\ 0.6837 & 0.0997 & 0.1024 & 0.1142 \\ 0.6089 & 0.1028 & 0.1351 & 0.1532 \end{pmatrix},$$

you find that the desired probability is $0.6089 + 0.1351 = 0.7440.$

6.3. This problem is an instance of the balls-and-bins model, as can be seen by imagining that the passengers inform the bus driver one by one of their destination before boarding the bus. The state is the current number of known stops. The p_{ij} are $p_{01} = 1$, $p_{i,i} = \frac{i}{7}$, and $p_{i,i+1} = 1 - \frac{i}{7}$ for $1 \le i \le 6$, $p_{77} = 1$ and $p_{ij} = 0$ otherwise. Calculating $\mathbf{P}^{10}$ gives $(p_{0k}^{(10)}) = (0.0000, 0.0000, 0.0069, 0.1014, 0.3794, 0.4073, 0.1049)$. As a sanity check, $7(1 - (\frac{6}{7})^{10}) = 5.502$ is the expected number of stops.

6.4. Let the indicator variable I_k be 1 if it is sunny k days from now and be 0 otherwise, given that it is cloudy today. Then $P(I_k = 1) = p_{CS}^{(k)}$. This gives $E\left(\sum_{k=1}^{7} I_k\right) = \sum_{k=1}^{7} E(I_k) = \sum_{k=1}^{7} p_{CS}^{(k)}$, and so the expected number of sunny days is 4.049.

6.9. Take a Markov chain with states $i = 0, 1, \ldots, 182$, where state i means that number 53 did not occur in the last i draws of the lottery. State 182 is absorbing with $p_{182,182} = 1$ and the other p_{ij} are $p_{i0} = \frac{5}{90}$, $p_{i,i+1} = \frac{85}{90}$ and $p_{ij} = 0$ otherwise. The element $p_{0,182}^{(n)}$ of the matrix product $\mathbf{P}^n$ gives the probability that within n draws there will be some window of 182 consecutive draws in which number 53 does not occur.

Note: the above solution method can be used as an approximation method for the following practical problem. For a Poisson process with rate λ, what is the probability that s of more events will occur in a time window of length w somewhere in a given time interval $(0, t)$? An example is Problem 5.35. The approximation approach is to divide the interval $(0, t)$ in $n = \frac{t}{\Delta}$ time slots of length Δ with $\lambda \Delta$ close to zero (say, 0.01) and then consider a sequence of n independent Bernoulli trials with success probability $p = \lambda \Delta$. An absorbing Markov chain can then be used to calculate the probability that there is somewhere a window of $s = \frac{w}{\Delta}$ consecutive successful slots.

6.10. A Markov chain with four states suffices. Take as state the number of filled glasses. State 0 is absorbing, $p_{10} = \frac{1}{3}$, $p_{12} = \frac{2}{3}$, $p_{21} = \frac{2}{3}$, $p_{23} = \frac{1}{3}$, and $p_{32} = 1$. The sought probability is $1 - p_{30}^{(10)} = 0.3660$.

6.13. The state of the Markov chain is described by the triple (i, r_1, r_2), where i denotes the number of smashed eggs, r_1 is the number of raw eggs picked by the guest, and r_2 is the number of raw eggs picked by the host of the game. The states satisfy $0 \leq i \leq 11$ and $r_1 + r_1 \leq 3$. The process starts in state $(0,0,0)$ and ends when one of the absorbing states $(i, 2, 0)$, $(i, 2, 1)$, $(i, 0, 2)$, or $(i, 1, 2)$ is reached. The guest loses the game if the game ends in a state $(i, 2, 0)$ or $(i, 2, 1)$ with i odd. In a non-absorbing state (i, r_1, r_2) with i even, the guest picks an egg and the process goes either to state $(i + 1, r_1 + 1, r_2)$ with probability $\frac{4-r_1-r_2}{12-i}$ or to state $(i + 1, r_1, r_2)$ with probability $1 - \frac{4-r_1-r_2}{12-i}$. In a non-absorbing state (i, r_1, r_2) with i odd, the host picks an egg and the process goes either to state $(i+1, r_1, r_2+1)$ with probability $\frac{4-r_1-r_2}{12-i}$ or to state $(i+1, r_1, r_2)$ with probability $1 - \frac{4-r_1-r_2}{12-i}$. This sets the matrix $\mathbf{P}$ of one-step transition probabilities. The probability that the guest will lose can be computed by calculating $\mathbf{P}^{11}$. It is easier to use a recursion to calculate the probability of the guest losing the game. For any state (i, r_1, r_2), let $p(i, r_1, r_2)$ be the probability that the guest will lose if the process starts in state (i, r_1, r_2). The goal is to find $p(0, 0, 0)$. This probability can be calculated by a recursion with the boundary conditions $p(i, 2, 0) = p(i, 2, 1) = 1$ and $p(i+1, 0, 2) = p(i+1, 1, 2) = 0$ for $i = 3$, 5, 7, 9, and 11. The recursion is

$$p(i, r_1, r_2) = \frac{4-r_1-r_2}{12-i}p(i+1, r_1+1, r_2) + \left(1 - \frac{4-r_1-r_2}{12-i}\right)p(i+1, r_1, r_2)$$

for $i = 0, 2, 4, 6, 8$ and 10, and

$$p(i, r_1, r_2) = \frac{4-r_1-r_2}{12-i}p(i+1, r_1, r_2+1) + \left(1 - \frac{4-r_1-r_2}{12-i}\right)p(i+1, r_1, r_2)$$

for $i = 1, 3, 5, 7, 9$ and 11. The recursive computations lead to the value $\frac{5}{9}$ for the probability that the guest of the show will lose the game. The expected value of the number of trials can be calculated as 6.86. Interestingly enough, the game turns out to be fair for the case of three raw eggs and nine boiled eggs, in which case the expected number of trials is 8.41.

6.17. The first thought might be to use a Markov chain with 16 states. However, a Markov chain with two states 0 and 1 suffices, where state 0 means that Linda and Bob are in different venues and state 1 means that they are in the same venue. The one-step-transition probability p_{01} is equal to $p_{01} = 2 \times 0.4 \times \left(0.6 \times \frac{1}{3}\right) + \left(0.6 \times \frac{2}{3}\right) \times \left(0.6 \times \frac{1}{3}\right) = 0.24$. Similarly,

$p_{11} = 0.4 \times 0.4 + 0.6 \times \left(0.6 \times \frac{1}{3} \right) = 0.28$. Further, $p_{00} = 1 - p_{01} = 0.76$ and $p_{10} = 1 - p_{11} = 0.72$. Solving the equations $\pi_0 = 0.76\pi_0 + 0.72\pi_1$ and $\pi_0 + \pi_1 = 1$ gives $\pi_0 = \frac{3}{4}$ and $\pi_1 = \frac{1}{4}$. The long-run fraction of weekends that Linda and Bob visit a same venue is $\pi_1 = \frac{1}{4}$.

6.18. Let state 1 correspond to the situation that the professor is driving to the office and has his driver's license with him, state 2 to the situation that the professor is driving to his office and has his driver's license at home, state 3 to the situation that the professor is driving to the office and has his driver's license at the office, state 4 to the situation that the professor is driving to home and has his driver's license with him, state 5 to the situation that the professor is driving to his home and has his driver's license at the office, and state 6 to the situation that the professor is driving to his home and has his driver's license at home. The process describing the state is a Markov chain with state space $I = \{1, 2, \ldots, 6\}$. The matrix of one-step transition probabilities is

from /to	1	2	3	4	5	6
1	0	0	0	0.5	0.5	0
2	0	0	0	0	0	1
3	0	0	0	0.5	0.5	0
4	0.75	0.25	0	0	0	0
5	0	0	1	0	0	0
6	0.75	0.25	0	0	0	0

Clearly, the Markov chain is periodic and has period 2. The equilibrium equations are $\pi_1 = 0.75\pi_4 + 0.75\pi_6$, $\pi_2 = 0.25\pi_4 + 0.25\pi_6$, $\pi_3 = \pi_5$, $\pi_4 = 0.50\pi_1 + 0.50\pi_3$, $\pi_5 = 0.50\pi_1 + 0.50\pi_3$, and $\pi_6 = \pi_2$. Solving these equations together with $\pi_1 + \cdots + \pi_6 = 1$ gives $\pi_1 = \pi_3 = \pi_4 = \pi_5 = 0.2143$, $\pi_2 = \pi_6 = 0.0714$. The long-run proportion of time the professor has his license with him is equal to $\pi_1 + \pi_4 = 0.4286$.

6.19. (a) The equilibrium probabilities are $\pi_{SS} = 0.6923$, $\pi_{SR} = \pi_{RS} = 0.0989$, $\pi_{RR} = 0.1099$. The long-run fraction of sunny days is $\pi_{SS} + \pi_{RS} = 0.7912$. **(b)** The long-average sales per day is $1\,000 \times 0.7912 + 500 \times 0.2088 = 895.60$ dollars. The standard deviations σ_1 and σ_2 are irrelevant for the long-run average sales.

Index

Printed in the United States
by Baker & Taylor Publisher Services